A Field Guide to Germs

Also by Wayne Biddle

Coming to Terms
(1981)

Barons of the Sky
(1991)

A Field Guide to Germs

Wayne Biddle

ANCHOR BOOKS
DOUBLEDAY
New York London Toronto Sydney Auckland

AN ANCHOR BOOK
PUBLISHED BY DOUBLEDAY
a division of Bantam Doubleday Dell Publishing Group, Inc.
1540 Broadway, New York, New York 10036

A Field Guide to Germs was originally published in hardcover by Henry Holt & Co. in 1995

Library of Congress Cataloging-in-Publication Data

Biddle, Wayne.
A field guide to germs / Wayne Biddle.— 1st Anchor Books Trade pbk. ed.
p. cm.
Previously published: New York: Henry Holt and Co., 1995
Includes bibliographical references and index.
1. Medical microbiology—Encyclopedias. 2. Medical microbiology—Popular works. I. Title.
QR46. B47 1996
616'.01—dc20 96-13621
CIP

ISBN 0-385-48426-7

Printed in the United States of America
First Anchor Books Trade Paperback Edition: August 1996

3 5 7 9 10 8 6 4

Nous sommes démocratisés et syphilisés.
—Baudelaire

• *Contents* •

· *Illustrations* ·

22. A Centers for Disease Control worker exposed to the Ebola virus. *Courtesy of National Library of Medicine.*
23. Rubella virus. *Courtesy of Centers for Disease Control.*
24. A Japanese woodcut, circa 1900, showing measles as an especially hirsute demon. *Courtesy of National Library of Medicine.*
25. Mumps virus. *Courtesy of Centers for Disease Control.*
26. "A Memory of the Mumps," *Harper's Weekly,* 1860. *Courtesy of National Library of Medicine.*
27. "Penicillin Cures Gonorrhea in 4 hours." *Courtesy of National Library of Medicine.*
28. "Gonorrhea, the Great Sterilizer." *Courtesy of National Library of Medicine.*
29. Malaria poster. *Courtesy of National Library of Medicine.*
30. Red blood cells infected with plasmodium. *Courtesy of Dr. Edward J. Bottone.*
31. "The War with Malaria," a Soviet educational poster. *Courtesy of National Library of Medicine.*
32. Board of Health quarantine notice, San Francisco, circa 1910. *Courtesy of National Library of Medicine.*
33. Elvis gets his polio shot. *Courtesy of March of Dimes Birth Defects Foundation.*
34. A cartoon published in 1826 shows a rabid dog causing panic in the streets of London. *Courtesy of National Library of Medicine.*
35. An antivaccination society illustration, London, 1802. *Courtesy of National Library of Medicine.*

· *Acknowledgments* ·

Researching a book like this one depends especially on the kindness of polymathic librarians. I wish to thank the staffs of the National Library of Medicine in Bethesda, Maryland, the Johns Hopkins University library in Baltimore, and the George Washington University library and Library of Congress in Washington, D.C., for their generosity. Mimi Harrison applied both her professional acumen and singular patience as the author's wife to the selection of illustrations, which profited from the guidance of Michael Harris in Washington and Edward Bottone at Mount Sinai Hospital in New York.

A
Field
Guide
to
Germs

An Egyptian stele, or stone tablet, from the eighteenth dynasty (1580–1350 B.C.*) depicting a man with a withered leg characteristic of polio.*

• *Introduction* •

Medical practice, like auto repair or the merger and acquisitions bar, is the bailiwick of high-pay specialists. The reasons for this are mostly obvious and good, yet there is strong agreement in the land today that ignorance among outsiders has been anything but blissful; that society's noblest interests are not always uppermost in every insider's mind; that it is possible to lose your shirt, or worse, on a bad afternoon. If this scary situation were confined to voluntary kinds of experience—air travel, say, or home remodeling—it might be acceptable as a lifestyle drawback. But health care is a universal parameter, from the Rwanda-Zaire border to Chevy Chase, a ubiquitous concern across every age, occupation, and class line. There will always be the lucky well and the unlucky sick, but the division should never be determined by privilege—intellectual or otherwise.[1]

A generation or so ago, the concept of first aid was still one familiar path whereby nonexperts rose a few inches above a state of total helplessness regarding medical matters, at least in emergencies. The shared experience of world wars and battlefield doctoring no doubt helped. Schools and community groups sponsored first-aid classes as they did art lessons. Accident manuals were common bookstore fare. Just a glance at an artifact so humble as the Boy Scout requirements for a first-aid merit badge thirty years ago reveals what now seems an extraordinary level of faith in the ability of ordinary people to deal with common predicaments: *Describe the causes and signs of shock; demonstrate what should be done for severe bleeding, poisoning by mouth, limb fractures; state some causes for unconsciousness; explain what to do for stomach pain, blisters, bruises, earache, choking, insect bites, heart attack, arm and leg cramps.* Though fixing these problems might entail relatively trivial skills, the underlying assumption, perfectly sound, was that even a twelve-year-old could be useful until the doctor arrived.

No new science has made such confidence obsolete, so what happened? Nowadays, of course, doctors do not arrive. It is the sick or injured who eventually arrive somewhere for attention. There are many reasons for this switch, not all of them as lofty as revolutions in medical technology. The way it seems is that the buyers of medical care lost pull to the sellers. Big-time capitalism may be partly to blame, with its penchant for giving consumers lots of choices but rarely any power other than the ability to reject inferior merchandise. Acquiring the services of a doctor is a singular market transaction, where the weakness of one party contrasts utterly with the stature of the other. "There are only two classes of mankind in the world—doctors and patients," Rudyard Kipling quipped. Myriad scientific advancements have widened the gulf between rich and poor nations, and similarly, access to medical expertise has grown more restricted for the poor.

Yet medicine has always been a cloistered subculture where unique responsibilities foster aloofness.[2] Add to this the fact that no profession ever had a better price maker than the Grim Reaper and there is obviously an ineluctable potential for abuse, the Hippocratic oath notwithstanding. During the

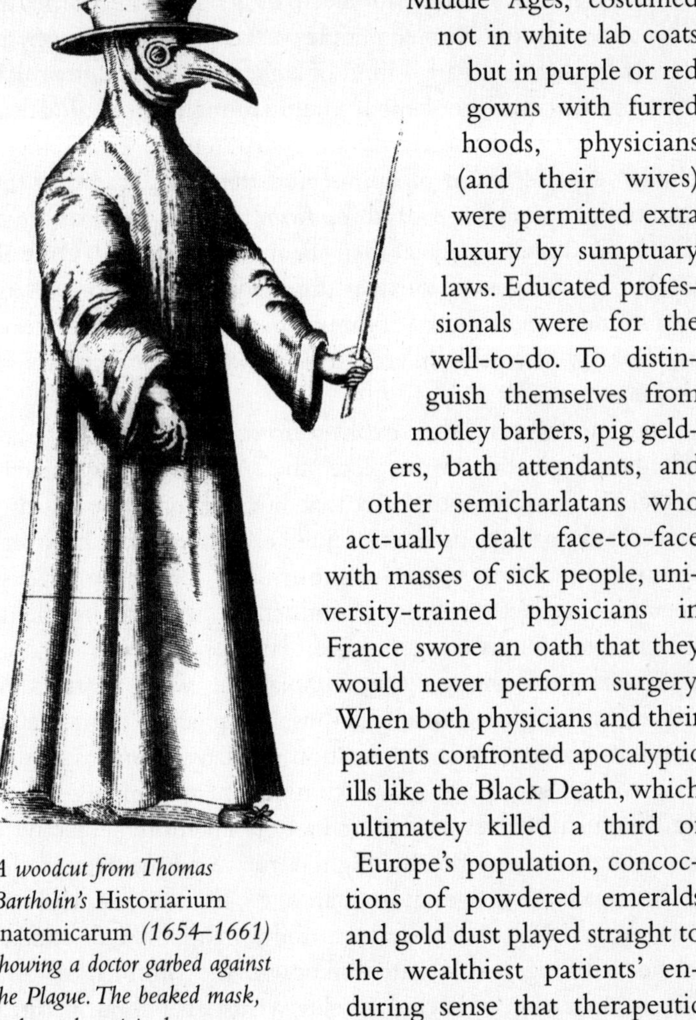

A woodcut from Thomas Bartholin's Historiarium anatomicarum *(1654–1661) showing a doctor garbed against the Plague. The beaked mask, perhaps the original nosegay, contained pleasant-smelling stuff to cover the reek of death.*

Middle Ages, costumed not in white lab coats but in purple or red gowns with furred hoods, physicians (and their wives) were permitted extra luxury by sumptuary laws. Educated professionals were for the well-to-do. To distinguish themselves from motley barbers, pig gelders, bath attendants, and other semicharlatans who act-ually dealt face-to-face with masses of sick people, university-trained physicians in France swore an oath that they would never perform surgery. When both physicians and their patients confronted apocalyptic ills like the Black Death, which ultimately killed a third of Europe's population, concoctions of powdered emeralds and gold dust played straight to the wealthiest patients' enduring sense that therapeutic value varies directly with cost.[3] ("If you want to be cured of I don't know what, take this herb of I don't know what name, apply it I don't know where, and you will be cured I don't know when," went one medieval piece of jocosity. Leonardo da Vinci, no lowbrow, warned people to "keep clear of the physicians, for their drugs are a kind of alchemy concerning which there are no fewer books than there are medicines.") Judaism taught that the physician was an instrument of Providence.[4] Universal prescription of opium into the first decades of the twentieth century for everything from the blues to fever must have helped make medical men seem angelic indeed. Only in the last several generations has the medical profession's practical foundation become very much deeper than the feel-good psychology of a self-confident *Herr Doktor's* taking control in vital emergencies.[5]

Like the folkways they evolved from, medical theories can still be dogmatic. Some, like the American Psychiatric Association's view until 1973 that homosexuality was a disease, are derived from rank prejudice. Others, like the overabundance of caesarian sections and coronary bypass operations, reflect nonscientific influences. Few establishments are capable of mustering more disdain for criticism of their orthodoxies, as though the complainant were obstructing Western triumphalism itself. Because the practice of medicine is not a science, but the application of many branches of science to the way people live (thinking of doctors as skilled craftsmen can be revelatory), it harbors far more subjectivity than state licensing boards might want to admit. Hence the common advice to "get more than one opinion," essentially to shop around. What remains the same for shoppers, of course, is the crushing fear of the unknown, the humiliation and depersonalization of having no clue whatsoever about nature's exigencies, the insult of seldom finding anyone who is glad to lay out a few basics.

Despite more than two decades of academic debate about a "new ethics" to bridge the social gap between doctor and patient, there is still a sizable reactionary force behind the notion that the power to heal arises at least partly from the gap itself. (We should keep in mind that the term "informed consent" was coined only in 1957, and then as a legal weapon.) According to this venerable way of thinking, most people are barely capable of assuming freedom and responsibility in medical matters when they are well. When they are sick, they want the "miracle, mystery and authority" of a paternalistic physician.[6] The so-called placebo effect—getting better after taking a fake pill—is thought to be evidence of this desire, but even bald-faced lying by a physician finds a time-honored place here, where "bad news" is felt to produce further loss of health.[7] A doctor's personal charisma and social clout are thus just as important, and sometimes more important, than medical knowledge, which is anyway imperfect.[8] "The presence of the doctor is the first part of the cure," says a French proverb. Better that the hopeful supplicant remain ignorant of medicine's limitations, or so the rather patronizing, self-serving argument goes.

In the United States, an unsavory congruence has developed between this condition of ignorance and the indecipherable excess of doctor bills, which has contributed directly to the nonpartisan demand for health-care reform. Staff parking lots at major hospitals filled with luxury cars do not send a reassuring message. As the perception of rip-off has risen, so has a general uneasiness about catching something. Economics is certainly a factor here, but probably a minor one compared to the slide of baby boomers into middle age, the decay of big cities into Third-World poverty, and the existence of a worldwide plague that continues to defy the power of research. Simultaneously, then, there are intimations of mortality among a vast cohort that never quite believed it would grow old, social policy failures in the very places where everyone was supposed to find the best opportunities, and an epoch of death-by-sex on the heels of the most bacchic sexual revolution in American history. Small wonder that a certain edginess about contagion has crept into daily encounters of all kinds.

It could be a lethal mistake, of course, to assume that the physician's craft is distillable beyond "first aid," or that "good practice" should pivot upon a patient's freedom of choice (what the ethicists refer to as "patient autonomy"). Medical diagnosis is not for hobbyists, any more than launching a shuttle is for space buffs. Doctors will continue to hold a tight monopoly on technical skills, even as their decisions are scrutinized through "managed care" systems.[9] Our enthralled society, moreover, is the least likely in the world to adjust doctors' socioeconomic status earthward. But signing a blank check is seldom a wise idea and never a healthy feeling, even for hypochondriacs. At a time when the nation finds itself spending $7,739 per family per year for medical care, a bit of straightforward knowledge about what this cash is aimed at may at least remove some of the attendant bitterness and head off further doctor bashing.

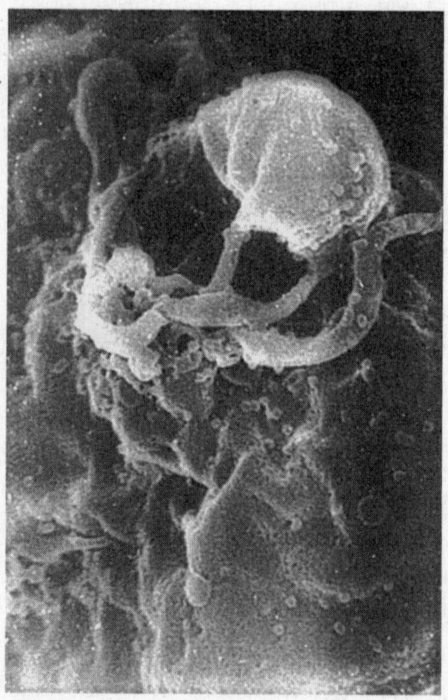

Fresh AIDS virus budding from the membranes of infected T-4 lymphocytes.

It may also help counter the phobias that so easily translate into political narrow-mindedness. Among the myriad tragedies

"Fear of disease causes the disease of fear." An early-nineteenth-century caricature about hypochondria and pseudoprotective measures against cholera.

of the AIDS epidemic has been the way it has played into the hands of a puritan element that would seek to proscribe certain expressions of human nature under any circumstances.[10] The horrific list of previously arcane diseases associated with the syndrome that have now become part of our lingua franca has helped fuel a pernicious "don't touch me" response that resonates with growing impatience with homeless people, expansive definitions of pornography, middle-class flight from nearly every urban public institution, and a Social Darwinist attitude toward poverty or unconventionality in general. If Michel Foucault, the French philosopher and historian of science, was correct in positing that the human body is a central metaphor for civilization's prevailing order, then the current debate about how to care for that body reflects deep social conflict.[11]

This distancing may be a primordial reflex—the Latin root of the word "infected," *infectio,* means tainted, spoiled, or polluted. Similar strains occurred during the American cholera epidemics of the nineteenth century, when only disreputable members of society seemed to catch the disease. The imagined contagiousness of their vice—be it poverty, Irishness, or some other moral failing—was as feared as their sickness.[12] Immigrants have traditionally been subject to such medicalized nativism.[13] Close analogies can also be found during the plague years of ancient and medieval history, when the stress of mass death altered the integrity of church and state. Beginning in 1348, thousands of European Jews were murdered for supposedly poisoning wells with Black Death. In 1630, at the start of an epidemic that would wipe out more than 30 percent of Venice's populace, laws were enacted to clear beggars from sacred quarters of the city. This was intended as an act of public atonement, not hygiene.[14]

Trying to make the best of quarantine on Ellis Island, New York.

Communal responses to disease are easier to organize when infec-

tion appears to be universal or random, rather than selective.[15] Thus, the repeated campaigns over the years to convince everyone that they are at risk for various venereal diseases, when the danger may be negligible outside specific circles.[16] Disease in itself has no social meaning, of course, being just the side effect of an invisible organism that is as integral a part of the planet's fauna as we are. It acquires meaning from the way it gives expression to our fears, which, as Aristotle said, arise from the anticipation of evil.[17]

The fact is that we live in a sea of countless microbes that outnumber us by many orders of magnitude.[18] Fortunately, most of them are harmless or beneficial. The pejorative "germ," from a Latin word meaning "sprout," is not a precise scientific term. The microscopic (and submicroscopic) creatures of concern here are mostly bacteria, living single-cell parasites that may cause disease through runaway infection of their host or the release of powerful poisons, and viruses, which are inert by themselves but have the nasty ability to pirate a host's metabolism. (Two categories of more complex organisms, fungi and protozoa, are of somewhat narrower interest.) The ones that can hurt us are called pathogens, from the Greek word for disease, *pathos.* Antibiotics slow the growth of pathogenic bacteria without zapping our own cells.[19] Since there are precious few antiviral drugs that will not also destroy the infected host, we depend on vaccines—usually made from safe pieces of the bug that nonetheless trigger our defense network—to combat viruses.

French chemist and microbiologist Louis Pasteur (1822–1895), known for gentleness toward small creatures, with his granddaughter.

It took thousands of years of very roughshod

German bacteriologist Robert Koch (1843–1910) and frau in Japan, 1903, where he was revered as a god.

medical practice and seat-of-the-pants science to arrive at these rubrics for microorganisms. Antoni van Leeuwenhoek (pronounced "layvenhook"), a Delft draper who became fascinated with lens making, reported on his observation of microorganisms in fresh water beginning in 1674.[20] He called them "animalcules." Though a Roman writer named Varro had speculated during the first century B.C. about whether people got sick after inhaling "tiny animals" (one can imagine his friends imploring him to lay off the grape), belief that such parasites came from nowhere—that is, generated spontaneously inside the body—prevailed well into the nineteenth century. A true germ theory of disease was spearheaded by the study of fermentation by French chemist Louis Pasteur (he was interested in the wine-making process, of course) beginning in 1857 and shown in a paper on wound infections published by the German bacteriologist Robert Koch in 1879. In general, the theory held that a pathogenic organism must be present in every case of the disease; that this organism could be grown or "cultured" outside the infected body; that an animal inoculated with this culture would get the same disease; and that the organism could be found again in the inoculated animal and recultured. Some pathogens may not meet all of these conditions, but they are the boilerplate for explaining scientifically how disease happens.[21]

For eons, we and the bugs have lived in a state of largely peaceful coexistence, modulated like everything else by natural selection. We depend on them, actually. Of course, a great deal of selection has already occurred—we are here because our ancestors somehow avoided deadly onslaughts. But different individuals and whole communities still show various degrees of resistance to infection by the relatively few organisms that are pathogenic. This can feed a scapegoat mentality for attaching blame when serious diseases occur, as when syphilis was called "the French disease" in Spain and "the Spanish disease" in France.[22]

Our defenses, which are formidable, are constantly adjusting. And so are the microbes, whose mutability is astonishing. The bugs have one big advantage: new generations may take only a few minutes to mature, compared to decades for our offspring (if we're lucky). Such interaction, generation after countless generation, creates a web of mutual adaptation that permits both host and parasite to survive. The most perfectly adapted parasites cause little or no damage, thus escaping the notice of everyone except scholars. If a disease organism kills lots of people very fast, however, it causes a crisis all around, including for the pathogen itself, because new hosts must be found often enough to keep its progeny going. In this sense, many of the most virulent germs can be thought of as gluttonous savages that are still in the early stages of biological adaptation (they may never be benign, just less disruptive). The time scale involved in the appearance on our horizon of a new pathogen is considerable: what we perceive as new or "emerging" pathogens are generally not straight out of nature's kitchen, but are rather existing microbes that have found us through changes in the environment or human ecology.[23] "We're boarding a train that's already in motion," said Luc Montagnier, the Pasteur Institute virologist who isolated the AIDS virus.[24]

Infections, like civilizations, rise and fall. Small comfort, yes, since the phenomenon cuts both ways and the balance of power is constantly changing. For example, when penicillin became widely available in the 1940s, most bacterial infections were treatable. By the 1960s, doctors were getting cocky about totally vanquishing infectious disease. But the evolutionary clock never stops ticking. Now, strains of common diseases have developed that are resistant to penicillin. Given enough time, new microbial strains are guaranteed to appear naturally on which current antibiotics have insufficient effect. This has already been noticed with certain relatively rare bacteria, but they may share their genes for resistance with other species that are more common. Most pathogenic bacteria are by now resistant to at least one antibiotic, a few to more than one. Once a resistant strain appears in a few scattered places, it can spread around the planet in a matter of months. Overuse of

An illustration from a poetry book, published in Paris by Auguste Marseille Barthélemy in 1851, showing a young suitor kneeling before Death disguised as a beautiful girl. That syphilis was frequently passed in the opposite direction would not have been fit to portray.

antibiotics, in livestock as well as in humans, speeds up the process by wiping out the unresistant competition. Many researchers now fear that the antibiotic spree in the decades after World War II will turn out to have been a terrible blunder, creating microbial freaks that assume unnatural dominance.

There are, to be sure, diseases that fly from victim to victim with the greatest of ease. The once common afflictions of childhood (common, at least, without vaccinations)—measles, chicken pox, sometimes deadly smallpox (now eradicated in nature), whooping cough, mumps—have at times been nearly as familiar as the sniffles. Happily, a single bout with them usually produces lifelong immunity. But if some of these same viruses were to invade an adult population with no previous exposure, we would see the same level of destruction as that which helped a handful of conquistadors to topple the Aztec and Inca empires, or that which decimated North American Indian tribes by as much as 50 to 90 percent after contact with Europeans.

Today, at least in developed countries, the most serious threat of infection comes not from virulent germs that exist outside our bodies, but from the myriad parasitic organisms we play normal host to (known rather daintily as our "resident flora"). They may live on our skin, in our mouth, in our throat or lungs, in our stomach or intestinal tract, and in our genitalia.[25] They are we. Since their ability to make us sick depends on the condition of our defenses, we are constantly warding off infection as a way of life. In a sense, we are all exposed, all the time.

When Baudelaire gazed upon the Paris barricades in 1848 and proclaimed "We are democratized and syphilized," his sentiment, albeit antirepublican, was an accurate estimation of the new order. "The rationality of life is identical to the rationality of that which threatens life," Foucault insisted 115 years later.[26] It is a rare individual who can embrace this tenet under duress, but the principle makes a good armature for personal attitude and public policy.

If there is a practical message here, it is that for most people who are able to buy and read this book, wickedly infectious microbes are rather hard to come by. The triumph of scientific medicine in this century has been to push calamitous encounters with pathogenic organisms away from bourgeois life. Deadly or disabling illness of this sort is more closely associated than ever with poverty, malnutrition, unconventional lifestyles (this is a loaded phrase, but it will have to do, the professional alternative being something ridiculous like "special sociobehavioral features"), sexual promiscuity (no moral judgment intended, just arithmetic), substance abuse, and other facts of human nature that tend to be ignored by polite circles whenever possible. Poverty has always put people in harm's way. Many of history's great epidemics have

IS THIS A TIME FOR SLEEP?

In an 1883 cartoon, Science sleeps on a New York dock as Cholera floats across the ocean from Europe.

shown strong "socioeconomic factors," as modern epidemiologists delicately say. But in the not-so-distant past the well-to-do were no better prepared to fight off a germ than the destitute. Today we have all the "miracles of modern medicine," as the catchphrase goes, but at a price that leaves much of the world's population in the dust.[27] In 1992, 33 percent of African-Americans and 29 percent of Hispanics lived below the federal poverty line, compared with 10 percent of whites—a situation that bears directly upon medical calculus in this country, which is unique among industrialized democracies for not guaranteeing universal health care. The United States ranks behind China, Mexico, and Albania in childhood vaccination rates. As America's inner cities experience Third-World poverty, they also encounter the forgotten-but-never-gone rationality of "that which threatens life."

The bourgeoisie should not grow smug. As existing infections are pushed to the margins, new ones will fill the space as evolution's imperative asserts itself. Nobody can escape the ecosystem.

This field guide is thus offered in the spirit of first aid, as a bulwark against phobia, and as a reminder that medicine is a social activity. There is room for humor and seriousness, history and science, politics and philosophy. Here, placed in a common context, are the top-ranked terms and germs (in prevalence, or power, or worry factor, or even literary interest) from among the billions of potentially unfriendly organisms out there, presented not with the quack promise of self-diagnosis, but with the absolute certainty that a little knowledge is always better than zip.

• adenovirus •

The adenoviruses are a family of more than forty viruses, identified by sequential letters and numbers, first found in adenoids (lymphoid tissue at the back of the pharynx) in the early 1950s, but definitely around for a long time before that. There are also adenoviruses that only infect animals. The ones that like people cause about 5 percent of all respiratory illnesses, from mild flulike symptoms to pneumonia, which are rarely fatal though of special concern among military recruits. Most children around the world have been infected with the more common adenoviruses by the time they reach school age. Some types are known to zero in on older children and adults in swimming pools and lakes.

Outbreaks of untrivial respiratory illnesses in new soldiers caused by adenoviruses have been a recognized problem since at least the Civil War. The term "acute respiratory disease" gained acronym status as ARD during World War II. Up to twenty cases a week per one hundred recruits could be expected, creating a significant drag on training. Just why some of the adenoviruses enjoy these young men so much, compared to civilians, is not clear. Populations of recruits have traditionally been subject to diseases usually seen in childhood, such as chicken pox, mumps, and rubella (German measles), perhaps because of stress, crowding, and the mixture of immune and nonimmune bodies from many different locales. Except during massive mobilizations, troops can be separated into small units when they first arrive, which helps to avert epidemics. Then again, many of these guys don't mind a few days in sick bay.

The ability of adenoviruses to infect our tonsils and adenoids for long periods of time implies that they have learned how to outwit our defenses. They do not fall into a latent state, like herpes viruses, but reproduce constantly at a snail's pace, changing the cells they enter in such a way that the immune system cannot tell the difference. Some of their tricks give them the power to produce tumors, but this has been observed only in unnatural hosts such as hamsters. Why adenoviruses cannot cause cancer in people is unknown.

• anthrax •

(*Bacillus anthracis*)

When the Lord visited "a very severe plague" upon Pharoah's cattle, as recounted in Exodus 9, it was surely anthrax. Tough *Bacillus anthracis* spores can persist for years in alluvial soil like that of the Nile valley, ravaging herds no matter whose side God is on—in this case, the Israelites were probably spared because they camped on sandy ground above the river's flood plain. Livestock dying in the grip of anthrax's gruesome spasms and convulsions would definitely seem cursed. (The lower Mississippi River valley is another well-documented haven for the disease—the first North American cases were reported among animals in Louisiana in the early 1700s.) Anthrax is known as a *zoonotic* disease (from the Greek *zo-*, "life," "animal," and *nosos,* "disease") because people catch it from animals—primarily cattle, sometimes sheep, horses, pigs, or goats—not from other people. Spores enter the human body by means of cuts (cutaneous anthrax, usually not fatal), the breath, or bad meat, then multiply and produce enough toxin within a week or so to injure surrounding cells and tissue, especially if inhaled (pulmonary anthrax, rare but quite deadly).

In mid–nineteenth-century England and Germany anthrax was called woolsorter's and ragpicker's disease, respectively, because these workers caught it from spores in fibers and hides. Robert Koch helped launch the science of bacteriology in 1876 by developing a method to grow pure anthrax cultures in the laboratory. When anthrax spread through sheep herds in France in 1877, Louis Pasteur began research that produced the first successful vaccination of livestock in 1881. Laws were enacted in the 1920s to require testing of shaving brushes that used horsehair or pig bristles. The largest outbreak of anthrax in the United States happened in 1957, when nine employees of a goat hair processing plant became ill after handling a contaminated shipment from Pakistan. Four of five patients with the pulmonary form of the disease died. In the 1970s, cases were traced to contact with souvenir drums from Haiti fitted with goatskin drumheads. An amateur weaver died after breathing spores in imported yarn.

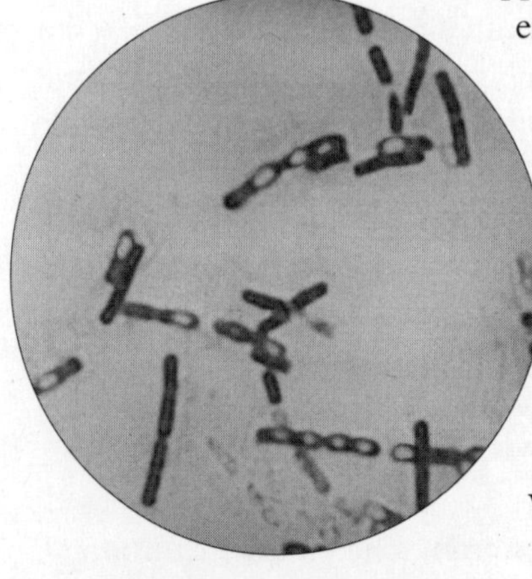

Anthrax cells, the first bacteria ever seen, magnified a thousand times.

Because the spores are so efficient, anthrax is considered a candidate for use in biological warfare. The U.S. Defense Department estimates that 50 percent of people who inhale between eight thousand and ten thousand spores will die. In the spring of 1979 an anthrax epidemic in Sverdlovsk, U.S.S.R., raised international concern about whether the outbreak was natural or due to activities at a military microbiology lab. The Biological Weapons Convention of 1972 prohibits work on offensive arms, although defensive development of vaccines is allowed. Soviet officials claimed at the time that victims developed gastrointestinal anthrax after eating contaminated meat and cutaneous anthrax from contact with infected livestock. In 1992 Russian President Boris Yeltsin said that the KGB had admitted a military cause. In 1994 an independent team of American and Russian experts determined that a windborne aerosol of anthrax spores from the military lab on April 2, 1979, produced the epidemic that killed sixty-six people out of seventy-seven cases, making it the largest documented outbreak ever of human inhalation anthrax. Why the release happened remains to be explained.

Today the risk of getting anthrax is negligible, at least in developed countries with modern animal husbandry and industrial hygiene. About 150,000 of the 200,000 troops sent to the Gulf War in 1991 were vaccinated against the disease. Living downwind from a military microbiology lab should perhaps be considered with the same caution as residing near a nuclear power plant.

• arboviruses •

The word "arbovirus" derives from the phrase "*ar*thropod-*bo*rne virus," which means viruses that propagate inside insects and other arthropods and reach us through bites. There are more than 520 known arboviruses, of which about a hundred cause disease in humans, usually with no apparent symptoms. But encephalitis, yellow fever, dengue fever, and a veritable gazeteer of exotic tropical fever malaises—Mayaro, Kyasanur, Bunyamwera, Marituba, Punta Toro, Candiru—give these microbes a deservedly bad reputation. Mosquitoes, ticks, and sandflies are the most common carriers, with various other flies and mites guilty to a lesser degree.

People are usually "dead-end" hosts for the arboviruses. That is, most of the germs don't need us for long-term survival—we just get in the way sometimes and then feel sick. Birds are much more crucial to arboviruses as long-term hosts. The exceptions of great significance are yellow fever, dengue, and chikungunya, where we serve as a vital link in their life cycle. Before the jet age, many of the arbovirus illnesses were encountered only by Indiana Jones types, but today they can turn up almost anywhere, at least as isolated cases. Epidemics are unlikely wherever the associated insects are kept under control.

• arenaviruses •

Arenaviruses—named after the Latin word for sand because of their granular interior—are a family of viruses first identified in 1933 during an encephalitis outbreak in St. Louis, Missouri; members of the tribe that have so far been identified sport names that would make a powerful magical incantation: Amapari, Junin, Lassa, Latino, Machupo, Parana, Pichinde, Tacaribe, Tamiami. Their natural hosts would spike the brew even more: bats, rats, and mice. Some of these viruses cause meningitis or various hemorrhagic fevers when humans come into contact with infected excreta. *Feh!*

They can be quite deadly. Argentinian hemorrhagic fever, caused by the Junin virus, discovered in 1953 near the Junin

River, may produce a 20 percent mortality rate as it disrupts capillaries and essentially causes its victims to bleed to death. The Bolivian brand (Machupo, also named after a river), first called *el typho negro,* the black typhus, can reach 30 percent mortality. It probably emerged via field mice of the genus *calomys* as jungles were cleared for agriculture during the 1950s. Lassa fever, which first came to the attention of Western scientists when American nurses were stricken in a Nigerian village of that name in 1969, topped 60 percent mortality in one hospital outbreak. Patients are therefore attended by doctors in biohazard moon suits—not a reassuring sight at bedside. Fortunately, most of these killers have remained confined to certain geographic areas, though cases have appeared elsewhere through the swift vagaries of modern travel. Direct person-to-person contagion is the exception, not the rule, so with proper surveillance these diseases should stay in the exotic category. Unfortunately, they occur in parts of the world where effective monitoring may be too much to expect of state medical agencies.

• bites •

As many as two million Americans are bitten by dogs every year, accounting for one percent of emergency room visits and more than $30 million in direct medical-care costs. Most nonrabid bites are provoked and most victims are males under twenty years old. Nearly a third of these wounds may become infected, regardless of first-aid measures like washing or applying salves. In general, bites to the hand are more serious and more frequently become infected than chomps to other areas. Infections are due mostly to germs in the dog's mouth, including pasteurella, staphylococci, and Weeksiella. Snakes account for another eight thousand punctured Americans annually. But human bites, including those self-inflicted by finger suckers, are far worse than animal bites. Many of the bacteria in human bite wounds may be resistant to penicillin. The bacteriology of Clenched-Fist Injuries (CFIs), where one bloke hits another in the mouth, is similar to that of bites and should be taken into consideration by Hollywood screenwriters. Vampires are already big money in Tinseltown, of course.

• *Bordetella pertussis* •
(whooping cough)

One of the planet's great ubiquitous organisms, the bacterium that causes whooping cough, *Bordetella pertussis,* cannot be avoided. It is everywhere and ferociously virulent, infecting 90 percent or more of susceptible individuals after exposure through close respiratory contact. It thus strikes early in life, with fatality rates highest among the youngest—a hundred years ago it was a major cause of infant mortality around the world and may still grimly reap wherever people are too poor or isolated to get immunized. It causes about 40 million cases and 400,000 deaths annually. A vaccine was widely available in the United States by the mid-1940s. Pertussis deaths began to decline for unknown reasons even before that decade, but local outbreaks still occur and increased during the late 1980s in the United States as a result of lax immunization programs. Between 1989 and 1991, there were 11,446 cases reported in the United States. The total of 6,503 cases in 1993 was the highest since 1967.

The terms "pertussis"(which means "intense cough" in Latin) and "whooping cough" first appeared in English in the seventeenth century. Before that the disease had been called "chin cough" or, in France, *quinta* (apparently referring to an interval of about five hours between coughing spasms) and *coqueluche* (also used for influenza). After incubating for one to two weeks, *Bordetella pertussis* invades the respiratory tract, causing a hacking cough. Over the next ten to fourteen days this progresses into a paroxysmal cough that may happen five to fifteen quick times in a row, followed by a high-pitched, crowing deep breath. Convalescence begins within a month, and most people recover after a seven-week illness. Pneumonia and brain damage are possible complications. Whooping cough was epidemic in the American colonies, particularly in South Carolina. *Bordetella pertussis* was not isolated until 1906, by the French bacteriologist Jules Bordet.

Fighting the germ requires diligent effort: in developed countries, children in the first six months of life usually receive three doses of vaccine (a combination of diphtheria, tetanus, and pertussis vaccines, or the DPT shot), with boosters at eighteen months and before starting school. Since 1940, this vaccine containing dead bacteria has cut whooping cough incidence by 99 percent in the United States, but it can produce serious side effects, including high fever, convulsions, and, in rare cases, brain damage and death. In 1988 the U.S. government established an $80 million fund to compensate children injured by the vaccine, prompting some 4,700 claims. Since then, the National Institutes of Health have spent $16 million to develop a safer vaccine, called "acellular" because it's derived from only part of a *B. pertussis* cell.

• *Borrelia burgdorferi* •
(Lyme disease)

Lyme disease, caused by the spirally coiled bacterium *Borrelia burgdorferi,* is today the most common insect-bite ailment in the United States and Europe. About ten thousand cases a year are reported in the United States (where it occurs almost everywhere, but especially in Connecticut, Massachusetts, Rhode Island,

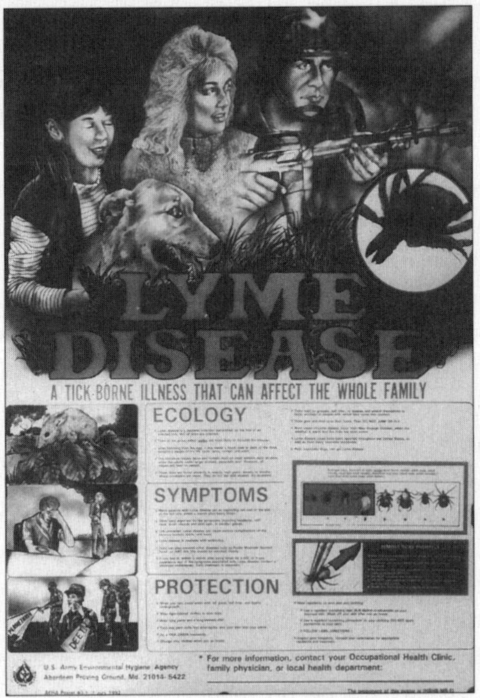

Delaware, New York, New Jersey, Pennsylvania, Wisconsin, and Maryland). Though this malady has caused unnecessary hysteria in some areas (enough to give it the nickname "borreliosis neurosis"), it provides a textbook example of how epidemics may seem to pop out of nowhere as a result of subtle changes in environment and human ecology.

By the early 1800s, vast tracts of land in the eastern United States had been denuded for firewood, lumber products, and farming. As virgin forest was replaced with an utterly different habitat of fields or low brush, the larger mammals that had thrived for eons all but disappeared, notably deer and their predators. By the mid-1800s, however, the process started to reverse as agriculture moved westward. Abandoned farmland did not turn back into the relatively open growth of great trees that characterizes wilderness, though, but into the scrubby woodlands choked with undergrowth that we know today. By 1980 there was four times more wooded acreage in the Northeast than there had been in 1860, but in biological terms it was a brand-new environment, a wide-open ecological niche. Deer began to return in the early 1900s (mostly without their old predators, which had been hunted to near extinction), and have reached record levels during the past decade. Also at record levels, of course, were the numbers of humans now living in close proximity to the deer. The conditions were perfect for a vicious circle of infection that shows no sign of being broken.

The first recognized outbreak occurred around the town of Old Lyme, Connecticut, in 1975 and 1976, though retrospective studies have identified a few apparent cases in the 1960s in Wisconsin and Massachusetts, and in forested parts of northern Europe in the 1920s. *Borrelia burgdorferi,* named after Willy Burgdorfer, the scientist who identified the species in 1982, reside harmlessly in white-footed mice (*Peromyscus leucopus*), which just happen to be the preferred feeding site for *Ixodes scapularis* tick larvae and nymphs—formerly called *I. dammini,* a suitable name, since they are damnably hard to see. (In California, *I. pacificus* ticks are the villains, and their menu involves lizards.) After the baby ticks eat mouse blood for about two days in the summer and four days the following spring, they become infected with *B. burgdorferi*. At the nymph stage they can infect humans. Adult ticks would rather suck deer blood, however, and it is at this stage that they may hop off a deer and soon chance upon humans who brush against infested vegetation. This cycle can be very efficient. *I. scapularis* were first discovered on deer carcasses in Ipswich, Massachusetts, in 1979. By 1986 the annual incidence of Lyme disease there exceeded 10 percent of the population, though this figure was probably inflated by overdiagnosis. There is no evidence of person-to-person contagion.

What can you do? Besides letting hunters shoot more deer, which is often controversial, or wearing tick repellant and long clothes in the spring and summer months, plus being sure to shower after a day outdoors (transmission of the disease does not happen unless the tick stays attached for thirty-six to forty-eight hours)—not much. In about half of people who catch Lyme, a blotchy red rash, often circular, will surround the bite within three to thirty days, and flulike symptoms will develop. But others may not notice anything abnormal for months.

In the northeastern United States, between 15 and 30 percent of all deer ticks are infected (or about half of all adult deer ticks). In most areas, only 1 to 3 percent of people who get bitten actually catch the disease, which is usually easy to treat with antibiotics if caught early. Unfortunately, current blood tests are poorly standardized and insensitive during the first weeks of infection. When the disease is diagnosed late in its course, it may be causing problems in many different organ systems, especially the skin, heart, nerves, and joints. Vaccines are being field-tested.

Some epidemic areas, like eastern Long Island, Block Island, Martha's Vineyard, and Nantucket require extra caution. Don't panic, shoppers: East Hampton's sidewalks are infested with neither mice nor deer.

• *Brucella* •

Most of us think of infectious disease as something caught from another person, but there are a few that can be picked up directly from animals. This kind of contagion is called *zoonosis,* a word that sounds threatening but should never keep anyone away from the zoo.

The genus of bacteria *Brucella,* named after David Bruce, a surgeon who first isolated them in 1887, has several members that mostly infect beasts with the illness now called brucellosis but can also cause trouble in humans. Because seminal research was done on the island of Malta on goatherds suffering from the disease by British army scientists in the latter half of the nineteenth century, it used to be known as Malta, Mediterranean, or Gibraltar fever. Brucellosis is largely confined in the United States to slaughterhouse workers but is sometimes connected to eating a fashionable gourmet item, fresh goat cheese. Aged cheese is okay.

Human brucellosis is caused by several different *Brucella* species: *B. abortus,* which resides in cattle, where it causes contagious abortion; *B. suis,* which prefers hogs and is more virulent when it attacks humans than its cow-rustling cousin; *B. melitensis,* which comes from goats and sheep and causes the most severe illness in humans; *B. rangiferi,* which likes reindeer and caribou; and *B. canis,* a dog lover. Many other animals, including horses, may carry *Brucella* bacteria. Humans are a dead end for these organisms, since there is no person-to-person contagion.

The action of *B. abortus* in causing cattle stillbirths is a classic example of the "specificity" of infections, that is, their adaptation to a specific niche. This organism needs certain vitamins and trace substances to grow, particularly one called erythritol, which is something like sugar. It just so happens that there is lots of erythritol in calf placenta, but not much elsewhere in a pregnant cow. So *B. abortus* thrives in the exact place where it can efficiently kill the animal's embryo. Happily, human placentas do not contain erythritol.

Brucellosis is rarely fatal, but the fevers (which go up and down, thus giving the disease another nickname, undulant fever), chills, weakness, and depression it brings are debilitating. Richard Nixon suffered from undulant fever as a teenager. *B. melitensis* is infamous for causing abortions in women, especially during the first trimester. In 1918 the researcher Alice Evans warned that *B. melitensis* was very much like the germ that caused contagious abortion in cattle. She caught brucellosis in 1922 after laboratory exposure and suffered from recurring fevers for twenty years, but gradually recovered and lived to the age of ninety-four.

In developed countries dairy herds are monitored for the bacterium's presence through tests of bulk milk. In any case, pasteurization zeroes the risk of contracting brucellosis. Though vaccines are available for animals, they are too toxic for humans.

• *Campylobacter* •

For many years, bacteria of the genus *Campylobacter* were thought to be related to the *Vibrio* germs of cholera infamy, which also produce intestinal illnesses. *Campylobacter* turns out to be much less dangerous, but far more widespread. It accounts for anywhere from 5 percent to 14 percent of all diarrhea in humans worldwide, making it the most common bacterial cause of this misery. It is one of the Big Four of the enteric—that is, pertaining to the intestines—pathogens, along with *Shigella, Salmonella,* and *Giardia lamblia*. Their basic effects are very much alike and require no elaboration here, please. They travel the ever-efficient fecal-oral highway via contaminated food and water whenever sanitation, hygiene, or cooking are, shall we say, inattentively carried out.

Two English veterinarians with strong stomachs were the first to identify *Campylobacter* microorganisms in 1909, while studying aborted cattle fetuses; they named the bacteria *Vibrio fetus.* In 1947 the same germs were found in a woman's blood under analogous circumstances. Not until 1972 did researchers realize that they were dealing with a new genus, which they named after the Greek for "curved rod," the bacteria's characteristic shape. Darwin would be pleased to know that campylobacters just happen to thrive at the body temperatures of their warm-blooded hosts, which include cattle, horses, poultry, and many popular pets, as well as people. "Thrive" may actually be too timid a word, since a single gram of an infected person's stool may contain a billion campylobacters—plenty to go around.

Wash your hands a lot if your puppy or kitten is sick.

• *Candida albicans* •

Candida albicans is a yeast, which is a kind of fungus, and it is definitely among us. It is not the sort of yeast that makes bread rise or beer brew or cider ferment, however. And it is not the awful stuff called Marmite that Brits love to smear on toast. Much like its bigger brethren growing on rotten tree trunks, *Candida albicans* is found naturally where the human body harbors the kind of decaying organic matter the fungus likes to eat—that is, in the mouth, genitals, or skin. (Don't feel insulted; even the most squeaky-clean aesthete has a lot in common with rotten tree trunks.) For the most part, the yeast becomes a problem only if our defenses weaken to the point where it grows too much, rather like a sandwich covered with mold as the preservatives wear out.

Diaper rash happens when *C. albicans* sets up shop on a baby's bottom; the immune system is not really a factor here, just the availability of a perfectly yummy environment. So-called yeast infections of the female genital tract are due to *C. albicans* finding its way there from normal homes on the skin or in the intestines, hardly ever by sexual means. A condition known as thrush—creamy white "cottage cheese" patches in the mouth or on the tongue that can be scraped off—which develops when *C. albicans* proliferates in that orifice, has gained notoriety as a sign of AIDS in young adults, but is also found in normal children.

Candidal jamborees can also appear around fingernails. None of these infections is serious in itself, yet most of us probably spurn the notion that we may sometimes resemble rotten tree trunks or funky sandwiches.

Baker's and brewer's yeast, by the way, is *Saccharomyces cerevisiae.* Food yeast à la Marmite, developed during the shortages of World War II, is *Candida utilis.*

• chikungunya and o'nyong-nyong viruses •

The chikungunya and o'nyong-nyong viruses are two arboviruses that cause diseases in Africa and Asia similar to dengue fever. O'nyong-nyong (the term means "weakening of the joints" in the Acholi dialect of Uganda) first appeared in Uganda in 1959, when it affected millions of people south to Malawi with fever, lymph-node swelling, and severe joint pain. It hitchhikes on *Anopheles* mosquitoes, also infamous for spreading malaria. Chikungunya, which means "that which bends up" in the language of the people of the Makonde Plateau in Tanzania, is a closely related virus, possibly a genetic precursor, that is endemic in East Africa. It rides in *Aedes aegypti* mosquitoes, which also transport dengue and yellow fever. There is no specific treatment for either chikungunya or o'nyong-nyong, so defense consists of killing or repelling the insects, sleeping under nets, and other safari-adventure movie stuff. Western physicians tend to call these viruses "novel," a perception that millions of Africans would no doubt find peculiar but not funny.

• Chinese restaurant syndrome •

You, too, may have suffered from substance abuse. Monosodium glutamate, aka MSG, the not-so-inscrutable ingredient in

Chinese food (at least the kind served to Occidentals), can cause chest pain, burning sensations, and the feeling that your face is blowing up like a balloon. The effects depend on how much MSG you eat and vary greatly from person to person. Perhaps the mischievous gods created this little curse as a reminder that MSG is manufactured by *Corynebacterium glutamicum,* a cousin of the diphtheria germ.

So, just say no.

• *Chlamydia* •

It sounds like something you don't want to get on you. *Chlamydia* bacteria have been bugging humankind for thousands of years, but as lifestyles have changed, the ailments they cause have shifted in significance. Scientists once thought that *Chlamydia* were viruses, because they cannot reproduce without attaching to other cells. The fact is that they possess all the biological features needed for independence, except the ability to generate their own energy. They are therefore known as "energy parasites," stealing their host's fuel to run their personal metabolism. Hostile takeover, indeed.

Three species of *Chlamydia* are of medical interest. *C. trachomatis* is the most important; it is passed around sexually and causes a variety of problems in human reproductive systems. It accounts for some four million infections annually in the United States, representing direct and indirect costs estimated at $2.4 billion. *C. psittacii* is carried by infected birds and gives us a kind of pneumonia. *C. pneumoniae* has been implicated in coronary heart disease. None of these diseases is usually fatal unless there are serious underlying diseases of other kinds. Still, they can be nasty.

C. trachomatis—the most prevalent of all sexually transmitted organisms, including the one that causes gonorrhea—

comes in two harmful strains. One produces infections of the eyes and/or genital areas. The other brings something called lymphogranuloma venereum, marked by swelling and ulceration of lymphatic tissue near the groin area, which has been fading away in the United States except among gay men. This disease is more common in tropical regions of Africa, Central America, and Asia, where it is of concern to visiting (i.e., invading) troops.

Strain number one's attack on the eyes, called trachoma, is now rare in the United States and Europe, but is still the leading infectious cause of blindness in hot, dry regions of the Third World. The main symptom is hard pustules on the inner surface of the eyelids. About five hundred million people are currently infected, resulting in six million to eight million cases of blindness. It was one of the earliest human diseases to be recognized as such, being described on papyrus in 1500 B.C. The name "trachoma," from a Greek word meaning "rough" (referring to the pustules), dates to A.D. 60. The disease spread out of the Holy Land with the Crusades and became known as "Egyptian" or "military" ophthalmia. Today it passes mostly among children and then to women caring for them. Clean water and control of eye-seeking flies usually put an end to it.

Strain number two has definitely been on the rise since the 1960s. The least bothersome thing it causes is conjunctivitis, inflammation of the rims of the eyes. The most tragic is infertility. When transmitted sexually, it can trigger a bevy of urinary-tract infections in men and women, pelvic inflammatory disease, and ectopic pregnancy. It may give a baby pneumonia during birth if the mother is infected. Like other genital pathogens, it is found most often among young, nonwhite, unmarried, poor people. In more than half of those who harbor this germ, no symptoms are apparent. Antibiotics can kill it, but using condoms is the best defense if you have lots of partners.

Recent research suggests that chronic infection with *C. pneumoniae* puts people at risk for atherosclerosis (hardening of the arteries, which causes more deaths in the United States than any other single disease) and heart attack. If this is proved conclusively, it could be a bonanza for public health, since the organism is vulnerable to antibiotics.

As for the pneumonia-like condition called psittacosis caused by *C. psittacii,* infection from casual exposure is exceedingly rare. Maybe you shouldn't feed the pigeons.

• cholera •

(*Vibrio cholerae*)

Cholera is one of the great social and political forces in human history. A filth disease that spreads via water and food contaminated by the *Vibrio cholerae* bacterium, cholera is now easy to prevent and treat, but remains a potent scourge wherever sanitation is not or cannot be maintained. Gabriel García Márquez chose well with his title *Love in the Time of Cholera,* there being few other kinds of experience more contradictory.

Historically, *V. cholerae*'s favorite breeding ground has been the Indian subcontinent, where clean water is still a pipedream. Twenty thousand pilgrims were killed by cholera at Hurdwar, a holy place in northern India, in 1784. Scientists count six distinct pandemics since 1817, during which the disease raged out of Bengal—probably, at first, because of expanded trade and the movement of British troops—and attacked ever more distant populations outside the country, only to slither back into India. During the second pandemic, which reached England in the early 1830s, John Snow began his classic investigation of the Broad Street pump, a public source of water drawn from a well near a cesspool in London's squalid Soho district. Snow, an unfashionable general practitioner, believed that cholera was waterborne, carried by the excreta of victims—in those days a brilliant but hard-to-prove guess. After he persuaded authorities to remove the pump's handle, thus shutting off the water supply, the neighborhood's outbreak waned. (In charge of cholera patients at the local hospital was an indefatigable thirty-three-year-old nurse named Florence Nightingale.) Snow was lucky not to be attacked by mobs who thought doctors were robbing the graves of cholera victims for anatomy lessons. On the Continent, French workers rioted because they feared they were being intentionally poisoned. Parisian ragpickers went on a rampage when the government ordered garbage collection as a preventive measure.

The disease returned to kill at least 61,000 people in Britain in 1848–49. In Paris, the same pandemic claimed 18,000 lives out of a population of 785,000, but riots by the working classes were averted when police relaxed the controls such as *cordons sanitaires* around poor neighborhoods. During this era, widespread death spawned some public-health innovations, but more often led to campaigns for moral reform.

The fifth pandemic, starting in 1881, brought rival German

Cholera Morbus
is about as sure to come as Summer is.
It comes suddenly and without warning — is Dangerous and often Fatal.
ARE YOU PREPARED for its coming?
If any of your family are attacked PROMPT action only may save life. For 46 YEARS ONE medicine has ALWAYS cured CHOLERA, CHOLERA MORBUS, DIARRHOEA, DYSENTERY and all SUMMER COMPLAINTS
CHILDREN can take it with perfect safety.
This medicine is
Perry Davis' Pain Killer.
To be on the safe side get some NOW and have it on hand.
For sale by all Druggists.
PERRY DAVIS & SON, Prop's. PROVIDENCE, R.I.

and French scientists to Egypt in order to discover the pathogen. The French team mistook ordinary blood platelets for cholera bacilli, then relaxed with swims in the sea, where one must have been a gulper. As the Frenchman later lay dying of cholera, the great German biologist Robert Koch, who had realized the French error and would soon identify *Vibrio cholerae,* comforted him by saying that he had indeed found the right germ.

An early-nineteenth-century German illustration of a man wearing protective gear against cholera—face mask, oversize shoes, a steaming pot on his head—as he pulls a wagon of earthly possessions.

All cases today are part of a seventh pandemic that started in Indonesia in 1961 with a new strain: *V. cholerae,* 01 biotype El Tor. Owing to the ubiquity of fast international travel and commerce, it is unlikely that the beast will ever retreat again to its Indian lair. Indeed, cholera can jump continents in a matter of days. Moreover, the El Tor vibrio can travel around the oceans in algae. Starting in several coastal cities of Peru after a Chinese ship dumped contaminated bilgewater, cholera reached epidemic proportions in South America in 1991, marking the first such outbreak there in a century. Victims soon appeared in Central America and Mexico. Over three years the disease struck at least

900,000 people and took more than 8,000 lives. Another newly identified strain of unusual virulence, *V. cholerae 0139,* swept across India and Bangladesh into Thailand in 1993, infecting hundreds of thousands of people and killing thousands. A case of 0139 infection popped up in California that same year in someone who had recently visited India. Cholera in North America is rare and isolated, most often occurring along the Gulf Coast, where raw oysters and undercooked crabs may transport endemic bacteria into unsuspecting gourmands. There have also been freakish imports from abroad, as in the episode of February 1992 when at least 31 of 356 passengers on a flight from Buenos Aires to Los Angeles were hit after eating contaminated seafood salad. In 1991 three Maryland revelers swallowed the bugs in frozen coconut milk from Thailand. Beware ceviche and piña coladas, gringos.

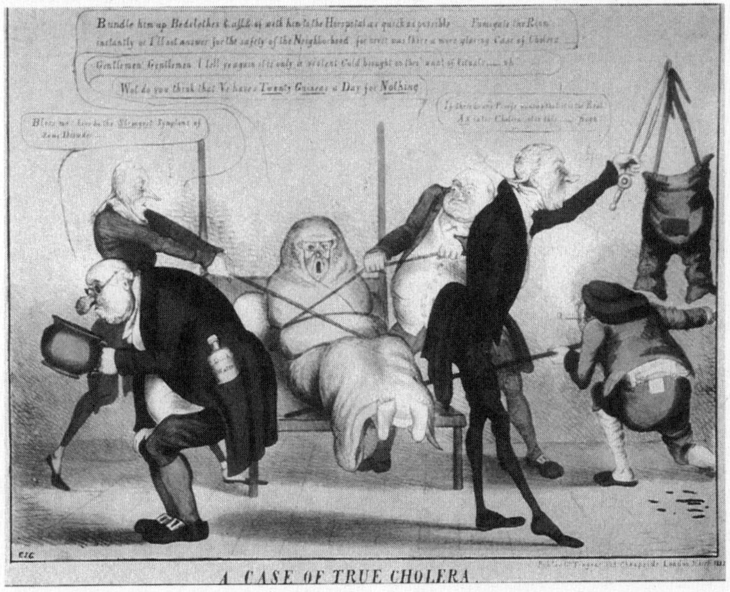

An illustration published in London in 1832 showing two physicians bundling up a cholera patient for removal while others inspect his dirty trousers and chamber pot.

Epidemic in Marseille, 1865. When nothing else works, build a big fire and dance around it like crazy.

One of the most grisly epidemics in modern times struck along the Rwanda-Zaire border in the summer of 1994. Earlier that year, famine-ravaged Somalia experienced a smaller outbreak believed to have spread from Ethiopia, whose government had suppressed reports of the disease. The chaotic circumstances of refugee camps create perfect conditions for spreading cholera through fecal-oral contagion. Humans are the only carriers of *V. cholerae,* so when hundreds of thousands crowd together in a prevalent region with no toilets or fresh water, the bacteria are virtually certain to start rampaging.

Yet, contrary to the hellish images from Rwanda, most cases of cholera around the world are mild or unnoticeable. Unless there is a gross source of microbes, like contaminated water or the clothes of victims, the disease is relatively hard to catch. Chlorinating drinking water prevents it. Salivary and stomach acids are a natural defense against *V. cholerae* until these substances are overwhelmed by sheer numbers of bacteria. Three quarters of all infections show no symptoms. But severe forms—aptly known for centuries as cholera morbus—where the victim has ingested millions of bacilli, for example, by drinking grossly polluted water, can be dramatically fatal when untreated. The bacteria multiply in the small intestine and produce a toxin that allows them to penetrate the protective mucous layer there, triggering a horrific outflow of bodily fluids and salts by the gallon. Sudden stomach cramps are followed by violent diarrhea, vomiting, and fever while ruptured capillaries turn the skin black and blue (whence comes the French phrase for transfixing fear, *une peur bleue*). Radical dehydration can bring death within a day through collapse of the circulatory system. Such gruesome effects contribute to cholera's fabled psychological impact on threatened populations.

If the victim's fluids and electrolytes are replaced in time, however, recovery is just as swift. Treatment could hardly be simpler: drinking, or taking by intravenous drip, large volumes of a solution of water, salts, and sugar. Ten thousand cases of Gatorade were shipped to the Rwandans as a quick approximation. UNICEF and other relief agencies maintain supplies of jiffy packets of salts and sugar for mixing with clean water. Antibiotics may be used to shorten the period of absorbing fluids and to ward off infection among the patient's immediate household, but are not distributed widely during epidemics because of the likelihood that resistive *Vibrio* strains would soon evolve. The Rwandan strain was determined early on to be resistant to tetracycline, a standard anticholera drug. The horror of cholera in refugee situations is thus compounded by the knowledge of how many lives—especially those of children, who may account for the most fatalities—could easily be saved.

For obvious reasons, cholera has traditionally been a disease of poor people living in sordid conditions. An exception was the 1970 outbreak in Jerusalem that struck the upper classes, many of whose victims were found to have highly alkaline stomach contents, through which acid-hating *V. cholerae* could pass unscathed. Terrible epidemics that struck New York, Boston, Philadelphia, and other American cities in the nineteenth century stigmatized slum-dwelling immigrants as the source of infection through their perceived intemperance and lack of hygiene. Previously unknown in the Western Hemisphere, cholera was considered alien to the bucolic New World experience—more the burden of poor Eastern Europe, say, where Poles cursed by saying "Cholera!"—meaning "Damn!" From immigrant neighborhoods it spread across the lowest strata of urban society, forming an indelible perceived relationship between disease and vice. The young nation's reaction to related issues of poverty, deviance from the mainstream, and socioeconomic assimilation is being played out again today on the AIDS front.

Cholera often serves as an alarm signaling a public-health system under stress. In 1994 an epidemic erupted in southern Russia and spread across Crimea into Ukraine. Former Soviet officials once proud of what had been accomplished under socialism are now faced with a poorly funded network of surveillance and treatment exacerbated by industrial pollution of water supplies.

Other members of the *Vibrio* genus may cause lesser diarrheal illness, blood poisoning, and wound infections—but none is as monstrous as *V. cholerae.* For example, *V. parahemolyticus* accounts for the majority of seafood poisoning cases in Japan during the summertime (usually from shrimp). It has also been reported in coastal American states such as Maryland. *V. alginolyticus* and *V. vulnificus* can infect wounds received in warm saltwater environments. The latter germ can also cause blood poisoning if swallowed, most often in people with weak immune systems or liver disease. *V. vulnificus* has turned up in raw Gulf Coast oysters, causing at least thirteen deaths since 1992. As the name suggests, all vibrios have a little waving tail that makes them frenetic swimmers (Koch compared them to swarming gnats), a feature that resonates with their penchant for ruining good water.

Kenne hora, as the Yiddish blessing goes: No cholera!

• *Clostridium* •

The genus *Clostridium* is a group of bacteria that, like large human families, has good and bad members. Named for the Greek word that means "spindle," these bacteria, which are cowardly themselves, produce rugged spores that can be found in dirt and dust all over the planet. Some are natural inhabitants of the intestinal tract, and thus the excreta, of many different mammals, including cattle, horses, and humans. Others are supreme opportunists, setting up house in a damaged host, where they produce poisonous toxins as they multiply. Because they are anaerobic and cannot thrive in fresh air, they love rotty places or sealed-off spaces. A few are therefore wicked causes of food poisoning and wound infection.

Clostridium botulinum is perhaps the family's blackest sheep; it is actually a gang itself, consisting of seven members, of which four may harm us. About two hundred years ago, scattered cases of muscle weakness and respiratory failure were linked to eating sausage and fish in Germany, Scandinavia, and Russia. The condition, which could quickly lead to death, became known as botulism from the Latin word for sausage, *botulus.* In 1897 a researcher named Van Ermengem found in a culprit ham a microscopic organism that would multiply only in oxygen-free containers. He went on to determine that it was not a parasite; that is, botulism was not an invasive, infectious disease, but the result of eating an incredibly powerful poison left in the ham by the microbe. This toxin could be destroyed with heat, he noted, and the organism would not produce it at all if the food was too salty.

Incidents of botulism rose with the practice of home canning, reaching a peak in the United States of about thirty-five cases per year in the 1930s—very few, but almost always the subject of publicity. The microbes release a gas, and swollen tin cans became the stuff of household nightmares. Since the thirties the number of incidents has steadily declined toward negligible, thanks to refrigeration and modern preservation techniques. In 1993, two restaurant outbreaks in Georgia and Texas were connected to canned cheese sauce and a dip containing baked potatoes, respectively.

Because *C. botulinum* spores are so ubiquitous in nature, it is conceivable that botulism could happen as an infection—in a dirty wound, say, or in the intestines—rather than from swallowing a toxin already formed in spoiled food. In fact, this turns out to be extremely uncommon in adults, yet for unknown reasons, botulism may take this path in infants under about eight months of age. This rare event may be one cause of sudden infant death syndrome. Honey has been implicated in some of these cases, so it is not recommended for babies. The toxin is so potent that even a lethal dose is much less than the amount necessary to trigger our defensive system to produce an antidote, so immunity does not follow such poisoning. If untreated, botulism can show a 65 percent fatality rate. Attentive medical care brings this down to less than 10 percent.

The tenacious spores of *C. botulinum* must be heated above 250°F (which is possible only in a pressure cooker) to guarantee destruction. They hate nitrites—hence the presence of that additive in hot dogs and other semicooked meat products. All in all, botulism is nowadays quite unusual, but anyone who "puts up" the bounty of their vegetable garden should be careful to follow home economics instructions. An experimental botulinum toxoid vaccine was given to about eight thousand military personnel during the Gulf War.

Anyone who is unable to put up with the realities of lost youth can call upon the botulism toxin for at least a temporary cure. For about $350, dermatologists in Los Angeles and New York—and maybe your town soon—will inject the stuff into eyebrow muscles, thus paralyzing them and smoothing away worry wrinkles. The treatment guarantees three to six months of looking blissful as a debutante. Because a tube of toxin loses its strength a few hours after being opened, Park Avenue doctors hold "botulism days" for all their eager patients. Enjoy.

C. perfringens is a more common, though far less hazardous, cause of food poisoning in the United States, especially in cooked beef and poultry. Ubiquitous in nature, its spores may survive the initial high temperatures of cooking, then germinate as the food cools and multiply if not properly refrigerated. If the food is served without enough reheating, live bacteria are then eaten by hungry human hosts, who get cramps and diarrhea eight to sixteen hours later. In 1993, two sizable outbreaks in Ohio and Virginia were traced to bad corned beef on St. Patrick's Day.

Far more threatening is the dirty work of another family member, *Clostridium tetani,* the tetanus bacterium. It, too, is dangerous because of the nasty toxin it releases, but not if you swallow some. Spores in soil may enter the body through almost any kind of wound—not just deep ones like the classic rusty-nail puncture, but even trivial splinter cuts—where they are then activated by decomposing tissue. Their toxin quickly finds its way through the nervous system into the spinal cord, making muscles contract (thus the nickname "lockjaw") and triggering spasms. A hideous grin, called *risus sardonicus,* may disfigure the face of untreated victims. Happily, the disease is totally preventable and now rare in developed countries, except among intravenous-drug abusers. In 1989–90, 117 cases were reported in the United States, about one quarter of which were fatal.

The peril of tetanus has been recognized at least since the time of Hippocrates, the fifth century B.C., and certainly in ancient Egypt, where dung was a favored wound medicine. It was especially notorious during military campaigns, when filthy lacerations and burns were the rule. After the Battle of Waterloo, the Scottish anatomist Sir Charles Bell wrote that "there are but two cases of living of all that had this affliction." Rather like botulism, tetanus was not understood until the late nineteenth century, but even then it was of major importance only on the battlefield. Later, mushrooming numbers of automobile accidents widened its swath. The great Japanese bacteriologist Shibasaburo Kitasato showed that the affliction was caused by the microbe's toxin. In order to find out how fast the poison traveled, he injected bacteria into the ends of mouse tails and then chopped the tails off as soon as an hour later, but too late for the mice.

Perhaps the most important discovery was that *C. tetani* toxin could be treated with chemicals like formaldehyde and made nonpoisonous—such a substance is called a *toxoid*—without ruining its ability to tweak our immune system into manufacturing a natural antitoxin. The toxoid was first used successfully on French cavalry horses after World War I. In addition, small amounts of toxin could be injected into animals, whose systems then produce an antitoxin that can be given to humans in emergencies (it does not bring the patient permanent immunity). Horses, especially sensitive to tetanus, are used for commercial production of the antitoxin. Thus, two routes of preemptive defense were established that have virtually wiped out the disease: long-term immunization with tetanus toxoid and/or injection with tetanus antitoxin within a day or two after injury. The first is nearly foolproof, the second not. Of the 117 cases of tetanus reported in the United States in 1989–90, 58 percent were in people over sixty years old, an age group less likely to have up-to-date immunization.

One heart-breaking indication of the enormous public-health gap between developed nations and the Third World is the existence of so-called neonatal tetanus. About 560,000 babies die every year in Asia and Africa because their umbilical cords are cut with filthy instruments or the stump is packed with dirt.

Clostridium difficile has recently been recognized as a cause of colitis associated with taking antibiotics. About 3 percent of healthy adults carry it around as part of the normal "flora" in their intestines, but it can also be picked up through fecal-oral contact, often in hospitals or nursing homes. Antibiotics change the natural balance of bacteria in the gastrointestinal tract, allowing *C. difficile* to multiply to the point where its toxin causes diarrhea problems.

At least one kind word can be said about clostridia. *C. pasteurianum* is one of the bacteria in agricultural soils that help plants acquire nitrogen. It is thus fundamental to the world's food production and perhaps more than makes up for the havoc wreaked by its fiendish relatives.

• Congo-Crimean/Rift Valley/California/St. Louis •

These far-flung locations are mainline stations on the arbovirus express. Different viruses may cause diseases with symptoms that are clinically indistinguishable. They can be told apart by delving into the body's biochemistry, but for the most part they are known for their geographical loyalties. Thus the melodiously malevolent chorus of locales associated with encephalitis and sundry fevers that in this case are carried to humans by insects.

A hemorrhagic fever—one accompanied by rash, bleeding gums and other mucous membranes, and sometimes internal bleeding—with a high mortality (15 to 20 percent) was first reported in rural areas of the southern U.S.S.R. during World War II. It appeared to come from the bite of ticks that infest rabbits. The culprit virus, technically in a category of arboviruses called *nairoviruses,* was isolated in the mid-1960s and named Crimean, but researchers in Africa found the same germ and called it Congo. So the scientific honor was split between the two regions, which would probably rather do without such an honor. In Africa, the virus has been found in all sorts of creatures besides ticks—cattle, goats, hedgehogs, and gnats—but in very few humans.

Rift Valley fever traditionally haunts the lands south of the Sahara Desert, bringing serious problems to humans and livestock. Caused by an arbovirus of the Bunyaviridae family, transmitted by *Aedes* and other kinds of mosquitoes, it was first noticed in 1930 as producing stillbirths in European breeds of cattle in Kenya and fever in veterinarians who investigated the problem and were thereby exposed to the animals' blood or viscera. After the completion of Egypt's Aswan Dam in the 1970s, which created vast new wetlands conducive to mosquito breeding, an epidemic hit some 18,000 people in the region and killed 598, also wiping out livestock herds. Epidemics related to dam construction happened in Mauritania, Senegal, and Madagascar during the next decade. Another outbreak occurred around Aswan in 1993, in which victims went blind after suffering fever, headache, and muscle pain. Livestock have been widely vaccinated in Egypt, but a vaccine for humans is not commercially available.

California and St. Louis are two of the five arboviruses of greatest significance to humans in the United States. They both cause encephalitis and are carried by mosquitoes (the former mainly by *Aedes triseriatus,* which breeds in tree holes and other containers, and the latter by several *Culex* types, which are the most common night-flying mosquitoes in North America).

Neither is confined to its namesake location. California encephalitis is seen most often in forest workers or other people, like campers, who spend lots of time in wooded areas. St. Louis encephalitis happens in irrigated farm country and urban or suburban areas where standing water provides breeding opportunities for *Culex* biters.

The other three important arboviruses cause eastern equine encephalitis, which is often fatal, western equine encephalitis, and Colorado tick fever, which is transmitted by the wood tick *Dermacentor andersoni.* Though unpredictable outbreaks occur from time to time, none of these diseases is common.

• coronaviruses •

Coronaviruses are some of the little devils behind the common cold among all ages worldwide—and are thus one of the inescapable annoyances of human life on earth. Named for their crown of club-shaped thorns, which can be seen with an electron microscope, they cause colds especially during the winter and spring (when another major group of runny-nose pests, the rhinoviruses, is at low ebb). Coronavirus-induced illnesses are sometimes fatal in lab animals, but not in humans, except perhaps at the extremes of vulnerability.

Coronaviruses need only about three days to multiply in the respiratory tract before their victim starts feeling miserable. On average, the cold lasts for a week—a few days shorter than a typical rhinovirus cold, but with more nasal congestion. Coronaviruses are remarkably good at reinfecting their hosts, which is one reason why vaccines remain elusive.

This germ is important not because of the severity of the disease it causes but because of the sheer number of clogged-up, grumpy humans it endlessly afflicts. What evolutionary purpose does it serve? Why does time only move forward?

• *Cryptosporidium* •

Americans are not used to getting sick from their tap water, but parasites of the genus *Cryptosporidium*—specifically the protozoan *Cryptosporidium parvum*—have recently caused trouble that used to be seen mainly in poor countries with unreliable water treatment or none at all.

Cryptosporidium organisms are about one sixtieth the size of the dust particles you can watch floating in a sunbeam. They have been found in 65 to 87 percent of surface water samples tested all over the United States. Cryptosporidium particles called oocysts, which have an egglike function, are passed in the feces of infected animals such as cattle and then find their way via physical contact or contaminated water into human intestines, where they hatch an infective spore that starts another reproductive cycle. Meanwhile, the person who swallows enough of them gets watery diarrhea and cramps for a week or more, an illness called cryptosporidiosis. Otherwise healthy people will soon cure themselves, but anybody with an impaired immune system is in for serious setbacks. There is no effective treatment besides replacing lost fluids.

Like all waterborne infectious diseases, cryptosporidiosis becomes a mass health hazard when water-treatment plants fail to do their job. In April 1993 some 300,000 Milwaukee residents developed severe diarrhea because city water-supply officials did not correct increased turbidity—that is, cloudiness of the drinking water that could be due to sand, silt, other nonorganic matter, or harmful organisms—during the spring runoff season. The outbreak, which required 4,400 hospitalizations, was the largest ever in an industrialized country.

In December 1993 more than a million residents of Washington, D.C., and northern Virginia were advised to boil drinking water after treatment workers at two U.S. Army

Corps of Engineers plants bungled the operation of filtration systems that draw water from the Potomac River. The public did not learn about the health hazard until the day after it started. Luckily, no illness was traced to this mishap.

Because cryptosporidia are not killed by routine chlorination (lab tests showed that the Milwaukee bug could live in Clorox), boiling water is the only practical defense against suspected contamination. Federal agencies recommend a rolling boil for one minute to wipe out all major waterborne protozoal and bacterial pathogens (this will also clear away hepatitis A, one of the more heat-resistant viruses that can spoil a good drink). If you live at an altitude above 6,562 feet, boil for three minutes.

• cytomegaloviruses •

Cytomegaloviruses, so-called because they cause enlargement of the cells (*cyto-* means "cell"), are catchall, do-all, everywhere, everybody viruses that are almost as much a part of human life as carbon and oxygen. Ninety percent of adults have probably been hit by one of these things, most without any symptoms, but a small minority contracting liver disease, mononucleosis, or respiratory problems. As is the case for many viruses, the reasons why they make some people sick and not others are unknown. They can sometimes be very damaging to newborns if the mother is infected for the first time during pregnancy—in countries where routine vaccination has defeated the rubella (German measles) virus as a cause of birth defects, cytomegalovirus infections may be the major viral source of such problems. More than 80 percent of toddlers—especially day-care kids—pick them up harmlessly as they pass around in saliva, tears, urine, and feces. People with unhealthy immune systems, particularly AIDS patients, may be direly affected, however. As members of the herpes group of viruses, they show the sneaky ability to reactivate after long periods of latency that so confounds attempts to fight them when necessary.

The cytomegalovirus particle is one of the biggest of all animal viruses, but at 200 billionths of a meter (200×10^{-9} meters, or 200 nanometers) in diameter it is still way beyond the reach of optical microscopes. (The single-celled paramecium that grade-school biology students often get to watch is more than a thousand times larger than a cytomegalovirus.) It is marvelously adapted for living in human beings and hates to be anywhere else. Years of searching for an infectious agent in certain cases of infant death led to isolation of the virus in 1956 from salivary-gland tissue (hence the medical nickname "salivary-gland virus").

Cytomegaloviruses often reside in blood vessels during infection, raising the fascinating possibility that they play a role in the cell changes that precede hardening of the arteries, or atherosclerosis. Perhaps far in the future we'll be vaccinated against heart attacks.

• dengue •

Some scholars maintain that "dengue" is a Spanish adaptation of the Swahili *Ki denga pepo,* used to describe a cramp that seemed to seize people like a spirit. Others point out that "dengue" is also a Spanish word meaning "affectation," an unkind reference to the mincing walk adopted by untreated victims of the disease to lessen the hellish pain in their joints. The English thus named it "Dandy fever." Dr. Benjamin Rush, the most eminent American physician of his day (and a signer of the Declaration of Independence), called it "break-bone fever" when it appeared in Philadelphia in 1780, because that's what it felt like. Today dengue occurs in the United States only when brought by travelers from areas where it is endemic, like the Caribbean islands, Central and South America, Africa, or Southeast Asia. In these locales, where billions of people are at risk, dengue is second only to malaria among insect-borne diseases.

Dengue fever is caused by an arbovirus transmitted mainly by bites of *Aedes aegypti* mosquitoes (which can also carry yellow fever and viral encephalitis), a variety of the common summertime pest that likes to breed in containers of still, clean water. The disease and the mosquito probably came out of Africa together with the slave trade and other commerce. While firmly established in Texas and Florida, American *A. aegypti* are not generally dangerous, because potential circles of dengue infection involving densely populated human settlements do not exist here anymore. An epidemic of more than a million cases occurred in the United States in 1922, but the last endemic outbreak of dengue fever happened in 1986 in south Texas. Only a dozen or so confirmed cases appear every year in the United States. Local outbreaks can occur at any time if someone brings the virus in from abroad and *A. aegypti* cooperates by biting the infected person, incubating the virus, and then biting another person. For the most part, though, dengue is found today only where governments either can't afford or don't care about mosquito control. Scattered cases of dengue began to appear among the thousands of American soldiers returning from Haiti in late 1994, raising concern that the virus could be introduced to wider populations.

Dengue is known for bringing on a sudden 104° temperature, nausea, vomiting, horrendous headache, a rash that appears after a twenty-four-hour pause in the fever, and long convalescence, but it is not deadly. Most cases are mild, treated with fluids and bed rest. There is a far more serious form of the disease, called dengue hemorrhagic fever, whose incidence has been increasing since first reported from Thailand and the Philippines in the mid-1950s. It hit 10,367 people in Bangkok between 1958 and 1963, killing 694. The dengue arbovirus comes in four models—they're all basically Chevrolets, so to speak, with different trim—and if you pick up two or more of them simultaneously, you get the really bad kind of dengue. In 1981 Havana was overwhelmed by an epidemic of dengue hemorrhagic fever that struck 344,000 out of a population of 2 million and killed 158. In 1982 more than 20 percent of New Delhi's 5.6 million people caught the disease. More than 3,100 cases occurred in Venezuela in 1990, with 73 deaths. The four types were probably once confined to separate geographic areas, but worldwide travel by infected people has mixed them together.

Another *Aedes* mosquito—*A. albopictus,* the "Asian tiger"—can also transport the dengue microbe. It arrived in Houston, Texas, for free with a shipment from Japan of used tires—a favorite breeding place—in 1985 and is now established in at least seventeen states. Though the aggressive tiger has not yet been found to carry dengue in the United States, another dangerous arbovirus, eastern equine encephalitis, has been found on *A. albopictus* around a tire dump in Florida.

• diphtheria •

(*Corynebacterium diphtheriae*)

George Washington died in his bed at his Mount Vernon plantation on December 15, 1799, because his throat was so swollen he couldn't breathe. It was a ghastly end for the Virginia gentleman, perhaps the result of infection with *Corynebacterium diphtheriae.* (The genus is so named because all its members are club-shaped, from the Greek *koryne,* meaning club.) Though physicians still debate whether George was done in by diphtheria, the disease's status in literary history is ensured by William Carlos Williams's 1933 story, "The Use of Force."

Although outbreaks of such deadly sore throats had been

noted sporadically since the time of Hippocrates, epidemics seem to have become common beginning in the sixteenth century. To prevent suffocation, tracheotomies were first used widely during an epidemic in Naples in 1610. In 1659 Cotton Mather described a "Malady of Bladders in the Windpipe" in Massachusetts that was especially vicious in children. Another New England epidemic started in New Hampshire in 1735, taking about a thousand lives, of which nine hundred were

Board of Health quarantine notice, San Francisco, circa 1910.

children. In 1826, during a period when the disease was prevalent in France, the physician Pierre-Fidèle Bretonneau aptly named it after the Greek word for a piece of leather, *dipthera,* referring to the tough, gray, mucousy membrane of dead cells that may block the victim's throat. The Spanish had once used an even more graphic term, calling it *garrotillo* after the executioner's implement, a string around the neck tightened by twisting a stick.

In the mid-nineteenth century the microbe responsible for diphtheria apparently became even more potent, because severe cases began to multiply sharply worldwide. In New York, where annual deaths averaged about 325 in 1872, fatalities jumped to more than 2,300 in 1875. *Corynebacterium diphtheriae* was identified in 1883, and by 1890 it was known that the bacterium produced a poison, called an exotoxin, which could be used in weakened form to prod the human immune system into producing a defensive material, called antitoxin. In this regard it was similar to tetanus (*see Clostridium*), whose mysteries were being unraveled at about the same time. Routine immunizations began in the 1920s, leading to near eradication of the disease wherever such programs are maintained. Not until 1951, however, was perhaps the most curious microbiological fact about diphtheria discovered: *C. diphtheriae* microbes that produce disease are themselves infected with viral parasites called bacteriophages. Actually, the gene necessary for producing diphtheria toxin comes from the bacteriophage, not from *C. diphtheriae.* Without it, *C. diphtheriae* would not be virulent.

Today, the tetanus-pertussis (whooping cough)-diphtheria shot (DPT) is among the first that children receive after birth. In 1993 not a single case of diphtheria was registered in the United States. In Russia, by contrast, with a public-health system under severe stress, recorded cases jumped from 3,900 in 1992 to 15,210 in 1993 to nearly 50,000 in 1994. The epidemic is most severe in cities on the Sea of Japan north of North Korea, where authorities launched a vaccination campaign at train stations, airports, and hotels.

C. diphtheriae bacteria thrive in dark, wet places like our upper respiratory tracts. Contagion usually occurs through close-up coughing or hand-to-mouth contact. Crowded, unhygienic living conditions help the microbes spread from person to person. Once established in a victim's tonsils, say, *C. diphtheriae* microbes are among the fastest microorganisms in producing symptoms of illness—not by racing throughout the body, but by the formation of powerful exotoxin. The poison may spread to damage tissues in the heart, kidneys, or nervous system. Death often comes from inflammation of the heart. Unimmunized sufferers must be given a quick jolt of antitoxin plus antibiotics and must be isolated under intensive care.

One of *Corynebacterium diphtheriae*'s cousins, *C. glutamicum,* has a much friendlier relationship with the human race. It is used to manufacture the all-purpose flavor enhancer monosodium glutamate, or MSG.

• dysentery •

"Dysentery" is a polite word coined by Hippocrates for the bloody runs. Its miseries have been described since the beginning of recorded history—Horus, son of the Egyptian gods Osiris and Isis, had dysentery. It has been of incalculable importance in military matters: from the Peloponnesian War through Napoleon's campaigns to the great sieges of this century, "campaign fever" has claimed more victims, through contaminated food and water, than has combat. William the Conqueror knew it well. Edward I and Henry V of England both succumbed to it, and the French mocked the dysenteric (though successful) English invaders at Crécy in 1346 as "barebottomed." George Washington suffered recurrent episodes of dysentery, exacerbated by hemorrhoids that sometimes required him to ride with pillows on his saddle. During the Civil War, the Union Army lost 81,360 soldiers to dysentery (and typhoid, with which it was then grouped), compared with 93,443 killed in combat. It can be caused by the bacterium *Shigella dysenteriae* (the genus named after the Japanese bacteriologist Kiyoshi Shigu, who discovered this particular dysentery bug in 1898) or an amoeba, *Entamoeba histolytica,* a slightly higher form of life. Human excrement is the primary source, with the amoebic disease prevalent where night soil is used as fertilizer. *E. histolytica* may travel through the bloodstream in prolonged cases to cause dangerous abscesses in the liver and brain.

Where living standards are high dysentery today may appear sporadically in overdeveloped suburbs, where wells have been dug too close to septic systems; in day-care centers and preschools, where kids do not wash their paws between potty and table; or aboard cruise ships, where the buffet is more dazzling than the galley deck. In poor countries, where public sanitation is lax, the disease can still rampage through wide populations. Since the 1960s, drug-resistant strains of *Shigella dysenteriae* have been cropping up around the world, sharply increasing the fatality rates for dysentery in poor countries. Lack of effective antibiotics is especially dangerous for children, one fifth of whom may die if they are untreated for *S. dysenteriae* infection (four fifths of children living in the tropics acquire bacillary dysentery before the age of five). In Ontario, Canada, 50 percent of all *Shigella* cases in 1990 were the result of organisms resistant to at least four antibiotics.

New *Shigella* strains have lately appeared in fast-food hamburgers, which may not get fried enough to kill the pests. The bacteria, and thus the burgers, are uniquely potent, given that fewer than two hundred cells can easily produce the disease in healthy adults. Yet the bacteria rarely cause severe illness in otherwise healthy babies, probably because breast milk contains defensive agents. It seems unlikely that fast-food restaurants will somehow incorporate this knowledge into their menus.

• *Enterococcus* •

Several *Streptococcus*-like bacteria that infect the human gastrointestinal tract used to be called enterococci. They are now classified in their own genus because they differ somewhat from proper strep germs. *Enterococcus faecium* and *E. faecalis* are especially worrisome, owing to their newly emerging super-resistance to antibiotics. They first appeared in New York City hospitals during the late 1980s and were held directly responsible for nineteen deaths out of a hundred infections (twenty-three other fatalities in the group had additional causes). Resistant strains soon turned up worldwide. By 1994, surveys indicated that 8 percent of all reported *Enterococcus*-caused illness was due to drug-resistant bacteria. The great fear among pathologists is that enterococci will transfer their resistance to other intestinal bugs, such as streptococci and staphylococci. The human gut is a perfect kitchen for this, making the dangerous outcome pretty much inevitable.

• enteroviruses •

The enteroviruses are a large mafia of viruses that inhabit the grand palazzi along the alimentary canal. The most infamous mobster causes polio. Others can trigger meningitis, encephalitis, respiratory illness, diarrhea, a severe form of conjunctivitis, and a dreary list of other afflictions, including the common cold. But all are known for inapparent infections in far more people than for producing active symptoms of disease. They are deceptive creatures, benignly tempered in most hosts but capable of vicious tantrums.

Besides the polioviruses, the group includes the Coxsackie viruses, named after the New York town where they were discovered during a 1948 polio epidemic; and the echoviruses, from the phrase "*e*nteric *c*ytopathogenic *h*uman *o*rphan viruses," which were thought of as orphans when first identified during the 1950s in the sense that their relationship to human disease was unknown. "Enteric" means isolated from the intestinal tract. A certain enterovirus can produce many different symptoms that may be similar to those of another enterovirus or even of a virus from a totally different group. Researchers have tried to cut through this natural fog with so-called Virus Watch Programs, where families deemed normal from a medical perspective are observed for years and regularly tested in an attempt to match the state of their health with the presence or absence of specific germs. In addition, the enteroviruses can be found so easily in sewage that much has been learned about how they infect various communities through testing of drainage systems (sewage treatment plants do not totally destroy viruses). Studies of this kind led to the startling knowledge that the ratio of inapparent (lacking symptoms) to clinical polio infections in New York City was more than one hundred to one in the 1940s.

Enteroviruses are tough little nuts. They are unfazed by antibiotics and most disinfectants. They can be frozen for years and come out fighting. They can stand being heated to 122°F. They are found everywhere at all times in the tropics and subtropical regions, but seem to prefer late summer or early fall in temperate latitudes. Chlorination will inactivate them, but the slightest presence of organic matter protects them, thus making water purification impossible in most settings. They crave human tissue—by the age of two, most children have had several harmless or mild bouts with enteroviruses—but fortunately for science they can be satiated sometimes with mice or monkeys. Highly communicable, only their habit of silent infection keeps them from being monsters capable of mass destruction.

The enteroviruses provide ample evidence that if an ecological niche opens up in nature, evolution will fill it. In this case, the usual expectation that poor children living in unsanitary conditions will suffer most is stood on its head. The frequency of severe outcomes of enterovirus infection, such as paralytic poliomyelitis, increases with age. The germs are usually spread by contaminated feces, so disadvantaged kids who lack good hygiene tend to get their first, immunizing infection early in life when they will probably experience minor malaise or no illness at all. As you climb the social ladder of cleanliness, first infection tends to be delayed to later childhood or young adulthood and beyond, when there is greater risk for paralytic disease, meningitis, and other serious problems. Franklin D. Roosevelt was a robust thirty-nine-year-old aristocrat when he was permanently paralyzed from the hips down in August 1921.

Outbreaks of hepatitis A associated with eating bad shellfish can be due to enteroviruses dumped into bays and estuaries with sewage. Clams, oysters, and mussels collect the viruses as they filter food out of the polluted water. Even if water is not dirty enough to break health regulations for bacteria content, it may contain viruses unperturbed by common treatment methods. Enteroviruses have been found along beaches miles from the nearest outflow pipe or offshore dump site. Since their favorite door into the human body is the mouth, swimmers are obviously at risk of infection. Testing for hazardous conditions is totally inadequate, making it wise for beachgoers to keep track of local construction projects.

A strong association exists between Coxsackie virus infection and insulin-dependent diabetes, based on thirty years of epidemiological studies. Coxsackie outbreaks tend to be followed by increased incidence of diabetes, perhaps because as the immune system fights off this virus, it inadvertently destroys insulin-secreting cells in the pancreas.

• Epstein-Barr virus •

"Our spirits rushed together at the touching of the lips," wrote Tennyson in 1842, not realizing that far more tangible stuff might be part of the flow. Unpoetic physicians call this "intimate oral contact," a pleasure more widely known as kissing. In the 1950s, a decade renowned for adult attempts to suppress intimate contact of any kind among those who desired it most—that is, teenagers—a researcher named

R. J. Hoagland observed that mononucleosis was spread by none other than intimate oral contact. Thus the powerfully erotic term "kissing disease" entered our vocabulary, disseminated by *Reader's Digest* and *Time* far more thoroughly than nature alone could ever accomplish. Mono was the sex scare of the fifties—tender evidence, perhaps, of how relatively innocent those days were.

The Epstein-Barr virus, named for the two scientists who discovered it in 1964, causes mononucleosis. It belongs to the primordial herpes family of viruses, whose members are infamous for their ability to persist or dwell latently for many years in their natural host, becoming active when the host's defenses weaken for any one among myriad reasons. Teenage cavemen no doubt got mono, though it was one of the least of their health concerns. It is most common in the ribald fifteen-to-twenty-five age group, probably because of more intense exposure, as the academics might say. In poor countries where hygiene is low, most children are infected by the time they turn six, but at this age the disease is mild or unnoticeable. Among the world's privileged classes, where saliva is abhorred, infection may more often cause serious problems—such as fever, jaundice, pharyngitis, or swelling of the spleen—because exposure tends to happen later in life. Thus were Yalies linked with the hazards of makeout parties in the late 1950s and early 1960s, when researchers at the medical school there found that three quarters of entering freshmen were at high risk for mono because they arrived without antibodies. Connecticut is still the only state that requires reporting of infectious mono cases—a tribute, perhaps, to the enduring vigilance of the Westport way of life. In general, however, no one yet knows why some young adults who carry the Epstein-Barr virus get mono and others don't, which is perhaps just as well.

In the mid-1980s, a mysterious long-term malady characterized by severe fatigue and other symptoms similar to mono became widely diagnosed as "chronic fatigue syndrome" (also known as "Yuppie disease" because it usually occurs in those between twenty and fifty years of age and has been popularized as an upper-middle-class phenomenon). Some sufferers show above-average levels of Epstein-Barr antibodies in their blood, while all other routine tests look normal. The virus's capacity, as a member of the herpes clan, to reactivate periodically over long spans of time could be behind this sometimes debilitating affliction, but it remains poorly understood. Fatigue is, after all, a subjective condition common to many diseases, psychological problems, and normal life in our high-pressure society. But chronic fatigue syndrome is no illusion.

Trying to control the virus is unrealistic, given the nature of intimate oral contact, and probably a bad idea, since exposure during childhood is better than later. Vaccination—of high-school students, say—would also be controversial, since there is a strong link between Epstein-Barr and cancer of the nose and throat in southern China, as well as a cancer known as Burkitt lymphoma in Africa and Asia, a connection that is not yet fully understood but might be activated by vaccines.

• *Escherichia coli* •

The rod-shaped bacteria known as *E. coli* are part of the normal "flora" of our intestines. The fact that they come in many different strains and are capable of reproducing at astronomical rates—one *E. coli* cell could multiply into a mass greater than the earth in three days if enough food were available—is the basis of their scientific usefulness and potential danger.

The vast majority of *E. coli* strains are harmless to most humans most of the time. For years they have served as the workhorses of biotechnology, providing fast-forward evolution for myriad types of experiments. However, the fact that they can multiply so quickly, doubling their population every two hours under the right conditions, means that they have the potential to make people, especially children and the elderly, sick. There are five groups of *E. coli* whose toxins cause diarrhea of various levels of severity. Some are the culprits behind traveler's tummy or "Montezuma's revenge." One country's flora are different from another's, so tourists may suffer a kind of poisoning until their immune systems adapt to newly imbibed bacteria. Others cause problems primarily in hospital nurseries. Still others produce a disease resembling mild cholera.

The worst *E. coli* bacterium, *E. coli* 0157:H, a mutation discovered in 1982 that has at least sixty-two subtypes, causes hemorrhagic colitis, an acute bloody diarrhea much like dysentery that can kill weak victims. Undercooked beef and raw milk are the microbe's favorite homes. It is also the leading cause of a kidney disease called hemolytic uremic syndrome. ECO157, as it is abbreviated, causes an estimated twenty thousand infections a year in the United States and is especially hard on children. Like many other strains of *E. coli,* this one's resistance to drugs is apparently due to overuse of antibiotics in livestock—it may live in the intestines of healthy cattle and then contaminate meat during slaughter. In 1991, twenty-seven people in Massachusetts were struck with ECO157 infection after drinking cider made from apples grown on trees fertilized with manure. In January 1993 contaminated fast-food hamburgers from the Jack-in-the-Box franchise triggered a multistate outbreak that killed at least four children. It struck not only those who ate the burgers but some who ate salads made with the same utensils used on the meat, as well as others who merely came into contact with people already infected. More than five hundred in the Seattle area got sick, with kidney failure occurring in at least twenty-nine. In 1994 the Department of Agriculture declared that raw ground beef found to be contaminated with ECO157 would be considered "adulterated" and prohibited from sale—the first time such drastic action was ever taken on raw food just for containing pathogenic bacteria. Five thousand samples of meat will be tested annually. Consumers can protect themselves at home by cooking ground beef to 160°F or until the inside is not pink and juices run clear.

ECO157 can also be waterborne, a fact discovered the hard way in July 1991, when eighty people were infected while swimming in an Oregon lake. In July 1993, 35,000 New Yorkers had to boil their tap water because ECO157 turned up despite chlorination.

Most *E. coli* are spread by contaminated food and water. So drink the beer and ask for "well done."

• Fort Bragg •

In the summer of 1942, young men unlucky enough to be training at this North Carolina Army base instead of peeping through beach house windows on Cape Cod came down with high fevers, headaches, rashes, and a bevy of other afflictions. They had picked up an organism of the genus *Leptospira,* meaning "thin coil," which was later named after their beloved home away from home. After the war, investigators traced the disease, leptospirosis, to exposure by swimming in freshwater ponds and streams contaminated by livestock urine.

Long before this outbreak the disease had been noticed among sewer workers, miners, farmers, and fish handlers. It was common in Japan, with mortality rates between 30 and 40 percent among thousands of victims every year. During the trench warfare of World War I, it plagued soldiers in Flanders. Rat piss was believed to be the culprit, but it is now known that myriad wild and domestic creatures—mice, dogs, bats, deer, foxes, raccoons, rabbits, pigs, goats, birds, frogs, snakes, fish, and on and on—play host to *Leptospira* and excrete zillions of microbes into the environment. The germs enter humans via direct skin contact. They are ubiquitous, but most of the forty to one hundred annual U.S. cases today occur in the Southeast, where water temperatures are optimal for the bacteria's reproduction.

When leptospirosis is severe it is called Weil's syndrome, after the German physician Adolph Weil, who first described it as "infectious jaundice" in 1886; it causes liver and kidney problems, meningitis, and vomiting. Antibiotics can control it, but the ecology of *Leptospira,* the number of species that can be hosts, makes prevention difficult.

• *Giardia lamblia* •

A protozoan parasite called *Giardia lamblia* will attach itself to your intestines with a little sucker, multiply like crazy, and make you fart. Have the gods no dignity? The creatures are found everywhere, but especially where sanitation is bad, since they pass around in food and water polluted with human or animal feces, particularly those of beavers and dogs.

In 1674 the pioneer Dutch microscopist, Antoni van Leeuwenhoek, observed *Giardia* in his own feces without knowing it: ". . . animacules a-moving very prettily; some of 'em a bit bigger, others a bit less, than a blood-globule; but all of one and the same make. Their bodies were somewhat longer than broad, and their belly, which was flatlike, furnisht with sundry little paws, wherewith they made such a stir in the clear medium and among the globules, that you might even fancy you saw a pissabed [louse] running up against a wall; and albeit they made a quick motion with their paws, yet for all that they made but slow progress."

Scientific jargon has clearly demeaned the language since Leeuwenhoek's day.

Like so many other pests that take the fecal-oral route into us, *Giardia* (named for a French biologist, Alfred M. Giard) tends to infect kids in day-care centers, gay men, and tourists from rich countries who are oblivious to the realities of life in poor countries. The United States is hardly free of the protozoans: *G. lamblia* has been found in 7 percent of stool samples tested nationwide, making it one of the most familiar intestinal infections. Fortunately, symptoms are usually mild or inapparent, but can include weeks of bad cramps, diarrhea, and considerable loss of weight. Antibiotics can achieve cure rates over 90 percent. The best way to avoid it is by not drinking untreated surface water.

• hantaviruses •

Hantaviruses are a genus of viruses carried by mice, rats, and voles and transmitted to us by breathing air contaminated by their droppings, urine, or saliva. The first of these viruses to be isolated, in 1976, was Hantaan virus, the one that causes a bleeding disease called Korean hemorrhagic fever, encountered during the Korean War; it was named after the Han River in South Korea. Hantaan had infected some 2,500 U.S. troops, killing between 5 and 10 percent of them, and was feared as a potential biological warfare agent. The related Seoul virus causes a similar but less severe fever and, probably because it is carried by common rats, is found everywhere. Puulmala virus, carried by voles, is found most often in Scandinavia and western Europe. They all cause their victims' capillaries to leak blood, leading to the failure of crucial organs before the immune system can counterattack.

In 1993 the genus gained fresh notoriety when a newly recognized member was blamed for the deaths of thirty-two people out of fifty-three cases, primarily in the Four Corners region of the southwestern United States. Press attention focused on sudden fatalities among young, otherwise healthy Navajo Indians who suffocated after their lungs rapidly filled up with plasma from damaged pulmonary blood vessels. Within a month after the first death, the Epidemic Intelligence Service of the Centers for Disease Control had identified a "new" hantavirus that they called the Muerta Canyon virus after a nearby geographic feature on the Navajo reservation—thereby offending the Navajos, who preferred to memorialize the canyon as the scene of a Spanish massacre (the pathogen was renamed Sin Nombre, or no-name virus). The American bug, which might have a fatality rate more than ten times higher than Hantaan, had been living for untold ages in deer mice (*Peromyscus maniculatus,* a ubiquitous North American field rodent), probably emerging from time to time to infect humans according to ecological whimsy. Traditional Navajo medicine men reported similar outbreaks during deer mouse infestations in 1918 and 1933. Indeed, Navajo legends recommended the burning of clothes soiled by mice.

Ninety-nine cases and fifty-one deaths from hantavirus pulmonary syndrome, as the disease is now known, had been confirmed in twenty-one states by the end of 1994, mostly in Arizona, New Mexico, and Colorado. After HIV and rabies, hantavirus would thus be the third most lethal virus ever found in the United States. Yet with closer study many of the recent cases have resisted being associated beyond all doubt with exposure to rodents or even to the virus itself, leading scientists to wonder whether still unknown factors are involved. A minority of the Four Corners cases, for example, were absolutely linked to Sin Nombre.

In January 1994 a twenty-two-year-old Rhode Island man died from the disease, though he had not traveled outside the Northeast for two months. Three possible sites of infection, all in New York state, were established, but not proved: an old warehouse in Queens that he had cleaned, a vacation house on Shelter Island, and his family home on Long Island.

In late 1994 the virus was confirmed in the case of a hiker in his mid-sixties who had been hospitalized in June 1993 after sleeping in a rodent-infested shelter along the Appalachian Trail in Virginia (he survived serious lung, kidney, and liver complications). There is no specific treatment for hantavirus infection. Scientists believe the germ will eventually be found throughout North America, threatening humans whenever rodent populations impinge on where we live.

• *Helicobacter pylori* •

Helicobacter pylori is the only known organism that likes to live in human stomachs, an acid-rich barrier against so many dangerous germs. Ignored for most of this century, *H. pylori* has been recognized only since 1993 as the prime cause of peptic ulcers, which afflict about 10 percent of Americans. Medical orthodoxy until then had deemed ulcers a noninfectious disease (to gather counterevidence, Barry Marshall, the physician who isolated the microbe in 1982, gulped a hefty dose of *H. pylori* and got an ulcer himself). *H. pylori* is more prevalent in older people and those living where sanitation is poor—more than three quarters of the Third-World population is infected, usually by the age of ten—but other than that there is no clear-cut reason why some individuals get stomach ulcers and others do not. What is new is the prescription of antibiotics for something always believed to be unrelated to germs. Mice have been successfully immunized, so a vaccine for humans is on the horizon.

• *Hemophilus* •

Bacteria in the *Hemophilus* genus are normal, harmless passengers in our upper respiratory passages. One kind, however, is a prime cause of meningitis in kids under three: *H. influenzae* type b, or Hib.

At the turn of the century, researchers believed that *H. influenzae* caused flu, hence the name. But after flu was shown to be a viral disease in the 1930s, this bacterium was relegated to the role of secondary infection. When the germ's epidemic potential was recognized in the 1960s and '70s, vaccine development accelerated. Today, children are vaccinated beginning at two months of age, assuming they live in the relatively few places on earth where immunization is affordable and performed conscientiously. Between 1987 and 1993 Hib disease declined by 95 percent in the United States because of vaccination.

• hepatitis •

The word "hepatitis" gets thrown around a lot in medical talk, because liver problems are so common. Many things can inflame the liver, often to the point of causing jaundice, the yellowing of the skin and tissues that is a telltale sign of liver disease, including alcohol, drugs, and other environmental chemicals of great category, as well as myriad microbes. Our modest purpose here will be served by zeroing in on the hepatitis viruses, which are unrelated but named after the first few letters of the alphabet: hepatitis A, B, C, D, and E.

Hippocrates was the first to note epidemics of jaundice, which certainly antedated his great acumen. John Dryden must have known of the disease, because he called jealousy "the jaundice of the soul" in 1687. The disease was often a factor in large military campaigns—during the Civil War, it struck at least seventy thousand soldiers. Until the middle of this century, hepatitis was pretty much synonymous with jaundice, which is actually just a warning that the liver can no longer cleanse the blood properly. By the beginning of World War II, investigators could differentiate between two types of hepatitis on the basis of how they spread—"infectious," via contaminated food or water, and "serum," via blood. It was not until the 1960s and '70s that the causative viruses were identified, partly with the help of a scandalous study of children at the Willowbrook State School for the Mentally Handicapped on Long Island, New York, where filthy living conditions had created a hepatitis infection rate near 100 percent.

Hepatitis A virus, the least dangerous, resembles the polio virus and consists of not much more than a bare strand of RNA genetic material in an icosahedral (twenty-sided) shell that reproduces only in the liver. It is known to infect only humans and a few other primates, mostly through contaminated feces. Hepatitis A is therefore a "filth disease" that tends to show up wherever sanitation and hygiene are poor. As with polio, people in poor countries tend to catch hepatitis A early in life, when symptoms are generally mild. Affluent travelers to such places may catch it as adults, however, and suffer badly. In the United States, where broad epidemics no longer occur, outbreaks are most common in day-care centers, among gay men, and in connection with polluted water or uncooked food, such as shellfish from contaminated water. Young infected children are almost never jaundiced, but are prime sources of contagion for adults, who usually show all the classic symptoms of nausea, vomiting, dark urine, and yellowish eyes and skin. Since 1971, when hepatitis A hit some fifty thousand Americans mostly in the fifteen-to-twenty-four age group, the disease has trended downward in numbers and upward in age. Today it shows strong "socioeconomic" ties, a polite way of saying it strikes hardest at destitute minorities. There is no commercially available vaccine yet.

Hepatitis B virus is a much more complex bullet than A and is natural only in humans, though it has close relatives that infect ducks, woodchucks, and squirrels. It can take as long as six months to incubate to the point of producing symptoms of disease, versus six weeks for hepatitis A. It passes from person to person in blood, saliva, and semen, which places it among venereal diseases. Before routine testing of the U.S. blood supply, thousands of deaths occurred every year from post-transfusion hepatitis of this kind. Moreover, the virus is extremely stable and can stay dangerous on needles, surgical tools, and even thorns or sharp stones. Hundreds of millions of people around the world are chronically infected and capable of giving the virus to others, unlike individuals who have had hepatitis A. In Asia and Africa, hepatitis B is commonly passed from mother to child during birth or in breast milk. Because the germ's long-term presence in the body often brings on liver cancer, it ranks as the world's most common viral cause of cancer. In the United States there are undoubtedly hundreds of thousands of such carriers, especially among gay men and injecting drug users. Between 1985 and 1993 the reported incidence of hepatitis B fell by 59 percent in the United States, probably because of increased condom use and safer needle-using practices among gay men and intravenous drug users due to AIDS awareness. A vaccine was licensed in 1982 after being tested with the help of New York City's gay community.

The relatively uncommon hepatitis C virus, which until 1989 was so mysterious that it was referred to as "non-A, non-B," is encountered mainly in the context of blood transfusions, drug abuse, and ingestion of contaminated water. It is related to the yellow fever virus and is a leading cause of chronic liver disease and cirrhosis. It is diagnosed by excluding A and B. Many questions about it remain to be answered, and there is no vaccine. Incidence of hepatitis C decreased by more than 50 percent in the United States between 1988 and 1993 as cases fell sharply among injecting drug users.

Hepatitis D is a handicapped sibling of the others, so to speak. It can only thrive in cells also infected with hepatitis B, whose coating material it borrows while it boosts the severity of disease. If B were a jet war plane, D would be the afterburner.

Finally, hepatitis E virus, which travels from host to host via fecal-oral contact and contamination of water rather like hepatitis A, is newly recognized as causing major epidemics in Asia, Africa, and Mexico. No vaccine is on the horizon.

Happily, most cases of hepatitis go away by themselves with "favorable outcomes," as the docs say, though illness can drag on for a month or two. After such protracted misery, most of us would feel jaundiced about life, indeed.

• herpes •

Herpes is one of the great virus crime families. There are more than ninety members that harass innumerable species, from us down to the lowliest fungi. They have been around for eons and are found in people everywhere, even in isolated primitive tribes. All have the capacity to survive for the entire lifetime of their host, hiding in nerve cells, sometimes producing symptoms of disease, often not. They have been implicated in causing cancer. Nothing can eradicate them from the body once they gain entry. Scary, yes, though generally not lethal.

The term "herpes," from a Greek word meaning "to creep," has been used for thousands of years. In popular speech it refers to just two of these thugs, *Herpes simplex* virus 1 and 2, abbreviated HSV-1 and HSV-2. The first usually causes cold sores or fever blisters, first described by a Roman doctor, Herodotus, around A.D. 100 as "herpetic eruptions which appear about the mouth at the crisis of simple fevers." Recurrent episodes can be brought on by emotional stress, ultraviolet energy in sunlight, physical exertion, other infections, and immune system disorders. Shakespeare wrote of this unsightly curse in *Romeo and Juliet*: "O'er ladies' lips, who straight on kisses dream/Which oft the angry Mab with blisters plagues,/Because their breaths with sweetmeats tainted are." The second type causes similar spots in the genital area and was not publicly reported until 1,600 years later, by a Frenchman named Astruc. HSV-1 and HSV-2 are very closely related creatures, about half of their genetic codes being identical. Either type can infect the same body sites under certain circumstances. The fact that one is a social nonissue and the other a scandal is a perfect example of how our reaction to disease reflects cultural values.

Thus, studies have been concocted that found 70 to 95 percent HSV-2 infection among prostitutes and, lo and behold, zero among nuns. Surveys also indicate that two thirds of people with either HSV-1 or HSV-2 show no apparent symptoms. Ergo . . . well, what exactly? A form of skin herpes named *Herpes gladiatorum* is found among college wrestlers and rugby players, who pick it up from mat burns and other repeated abrasions. So how about tarts who wrestle? Rugger-bugger nuns?

A drug called acyclovir is currently the best defense against HSV outbreaks. It is not a cure, but it reduces the number and duration of attacks in people who experience them. Debate about whether to make acyclovir available without a prescription has focused on the likelihood that resistant strains of HSV would spread more rapidly than otherwise. This technical fear may be overwhelmed by sheer demand: some 55 million Americans carry HSV-2, 11 million suffer periodic outbreaks, and the numbers are rising by hundreds of thousands every year.

• HIV •

(human immunodeficiency virus)

Before descending into a welter of detail, let it be said from the start that prevention is still the only reliable way to fight the HIV monster.

AIDS, *a*cquired *i*mmuno*d*eficiency *s*yndrome ("syndrome" means the occurrence together of a characteristic group or pattern of symptoms), has been widely called a modern Black Death. Would that this were so, because then the cure would be easy. The comparison is valid, however, when one considers the state of fear and helplessness that AIDS has engendered since it was recognized in the early 1980s. The tenth International Conference on AIDS in 1994 was notable for neither optimism nor pessimism, but lots of realism. AIDS is simply going to be around for a long, long time. Scientists are now reemphasizing basic research over clinical advances—not exactly going back to square one, but close. After a decade of frenetic investigation at the forefront of virology, it is apparent that HIV and its effects on the immune system are not yet understood deeply enough to lead to dependable treatments.

A disease as catastrophic as AIDS takes on myriad brands of sociopolitical baggage, clouding realities that are horrid enough on their own. The World Health Organization estimates that 17 million people are infected with HIV and 4 million have AIDS. More than 90 percent of AIDS cases occur in Third-World countries. The largest number of cases, estimated at more than 2.5 million, is in sub-Saharan Africa, where there are more than 10 million HIV-infected adults. In the United States, AIDS is still overwhelmingly an affliction of homosexual and bisexual men (47 percent of reported cases in 1993) and injecting drug users (28 percent of reported cases in 1993). Among racial and ethnic groups, African-Americans and Hispanics account for both the majority of reported cases (54 percent in 1993) and the greatest annual increases. Heterosexual transmission, 42 percent of which was related to contact with an injecting drug user, occurred predominantly in these groups (78 percent of men and 74 percent of women in 1993). Factoring in age reveals an even grimmer picture: in 1993 minorities accounted for 51 percent of reported cases

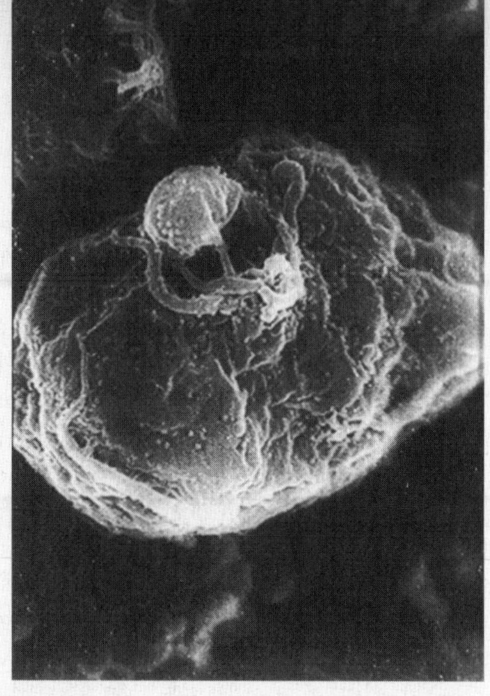

AIDS virus bursting from the membranes of infected T-4 lymphocytes.

among adult and adolescent males and 75 percent of cases among such females. Among children (younger than thirteen years old), 84 percent of AIDS cases were minorities. The AIDS rate for black women was about 15 times greater than for white women and almost five times greater for black men than for white men. In 1991, AIDS was the leading cause of death among black and Hispanic males aged 25 to 44 and the third leading cause for women in the same age group (in the overall American population, the ten leading causes of death, in decreasing order, are heart disease, cancer, stroke, lung disease, accidents, pneumonia and influenza, diabetes, AIDS, suicide, and homicide).

These numbers are chronic dynamite because communal responses to epidemics are always easier to organize and sustain when the disease is perceived as being universal rather than marginal. Throughout history, diseases that disproportionately strike poor people (or people practicing what the larger society defines as vice) have been stigmatized to an extent that ultimately proves more damaging to society than the maladies themselves. Top federal officials in the United States originally opposed efforts to confront AIDS, apparently because of their bigotry toward gay men and drug addicts. Former Surgeon General C. Everett Koop's comments in 1991, reported in Lederberg et al., *Emerging Infections,* are often cited as testimony to this scandal: "Even though the Centers for Disease Control commissioned the first AIDS task force as early as June 1981, I, as Surgeon General, was not allowed to speak about AIDS publicly until the second Reagan term. Whenever I spoke on a health issue at a press conference or a network morning TV show, the government public affairs people told the media in advance that I would not answer questions on AIDS, and I was not to be asked any questions on the subject. I have never understood why these peculiar restraints were placed on me. And although I have sought the explanation, I still don't know the answer." Despite the fact that AIDS research is now well institutionalized in the medical establishment, continued opposition to sex education, condom distribution in public schools, and clean-needle programs shows the reach of such "peculiar restraints," which can amount to a de facto death sentence.

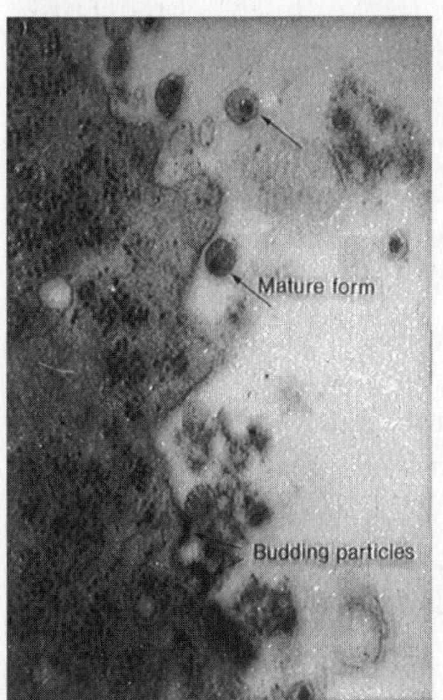

AIDS virus particles found in a hemophilic patient.

Unlike the Black Death, which tended to come and go, AIDS may belong in the same realm as cancer or heart disease. That is, for some but not all populations it is an endemic killer. Or it may be similar to tuberculosis in the last century. It differs from all three points of comparison because it is totally preventable, communicable but not contagious. Of course, as a disease that is largely venereal, like syphilis, AIDS is subject to the debased mentality of divine retribution.

The literature on AIDS is already vast and perhaps the fastest-growing body of medical research. Interested lay readers can stay abreast with the journals *Science* and *Nature*. Briefly, HIV (*h*uman *i*mmunodeficiency *v*irus) is the third of five human *retroviruses* that have been discovered since 1980 (there are undoubtedly more). A retrovirus is one containing an enzyme, *r*everse *t*ranscriptase, that converts viral RNA into a DNA copy that becomes part of the host cell's DNA. Before that, this class of pathogens had been connected only with various diseases in animals. HIV, which actually refers to two closely related viruses that cause AIDS in separate geographical regions, is part of a class of retroviruses known as *lentiviruses* traditionally associated with chronic arthritis and anemia. (Lentiviruses are retroviruses that cause slowly progressive, often fatal diseases.) HIV's unprecedented, tragic specialty is that it likes to attach itself to the very cells that turn on and amplify our immune system—the so-called T lymphocytes, specifically those with a protein called CD4 on their surface, which is the actual hookup point for HIV. Other lentiviruses do the same thing in cows and cats.

Once inside the T cell, the virus releases its genetic template (RNA) along with a chemical that allows it to be transcribed into the cell's own DNA. All offspring of the altered T cell thus contain the virus's genetic code. The T cell also may become a factory for new infectious HIVs, which destroy it as they burst out. The regulation of this life (or death) cycle is not fully understood, but one of the hallmarks of the process is the extremely long time—from two to fifteen years—that may pass between initial infection and onset of the syndrome (latency). Moreover, HIV constantly undergoes rapid genetic mutations in each individual victim, so that the immune system, or what's left of it, is always struggling to hit the right target.

The immune system fights a long, ferocious, but ultimately losing battle against the AIDS virus. In late stages of the infection, victims lose and replace about 2 billion CD4 lymphocyte cells a day, while new virus particles appear at a rate between 100 million and 680 million a day. Other viral diseases, such as the flu or hepatitis, may also trigger warfare of such scope, but for a relatively brief time.

This is what makes the development of a single "magic bullet" vaccine so unlikely. No one knows if a vaccine made from one strain will protect against others. Thirteen experimental AIDS vaccines have been given to more than 1,500 volunteers in the United States since 1988, but federal health officials have so far rejected large-scale trials due to lack of confidence in their potency. The World Health Organization nonetheless plans to test two such vaccines, probably in Thailand and Brazil, in a desperate attempt to find one that provides even a low degree of protection. If the gods had wanted to design a perfect time bomb, it would look like HIV, which perhaps helps to explain why some people hang on to the myth that the virus was created by biological warfare experts.

Where *did* it come from? The short answer is that no one knows. The fact that there are similar lentiviruses in animals suggests a plausible source, but how HIV hopped into humans is unclear; monkey bite is as good a guess as any. (The earliest documented case of HIV infection was found in a 1959-vintage blood sample from central Africa.) Epidemiological evidence suggests that it traveled along the Mombasa-Kinshasa highway in Africa and thence to the United States during the mid-1970s, where it was first transmitted via anal intercourse among highly promiscuous (that is, hundreds of different partners a year) gay men in San Francisco and New York. By mid-1981 the Centers for Disease Control began to notice in young gay men rare diseases—pneumocystis carinii pneumonia and Kaposi's sarcoma—that had previously occurred only in people with damaged immune systems. To other industrialized nations AIDS then became "the American disease." Because the virus passes easily in blood, intravenous-drug users were the next vulnerable pool, along with the much smaller number of hemophiliacs and others who received contaminated transfusions. Since the virus can also go from person to person through vaginal intercourse (the primary route in Africa), the coterie of sexual partners of these victims was soon infected, too.

And so on and so on. In the vicious AIDS circle, you essentially experience all of your lover's old flames. Pathogens adapted for this kind of infection tend to become permanent parts of the human condition, regardless of whether cures exist. The World Health Organization has estimated that worldwide, forty million people could be infected by the year 2000. Still, the authors of the most comprehensive survey ever of American sexual habits, released in 1994 under the auspices of the University of Chicago, concluded that HIV infection and AIDS are unlikely to become epidemic in the general U.S. population because Americans tend to have sex with partners from similar social milieus. With few "bridges" between the marginal and the mainstream, the disease will not spread easily from those groups already hardest hit. Besides, more than 80 percent of American adults have only one sexual partner, or none at all, in a year's time, the survey found.

The risk of getting AIDS from casual, household, or workplace contact is still put at near zero, even from personal items such as toothbrushes, razors, clippers, and towels. Ditto for blood-carrying insects like ticks or mosquitoes. In addition to blood, the virus can be found in semen, breast milk, vaginal secretions, and saliva, though there is no evidence that AIDS can be caught from an infected person's spit, as in coughing or sneezing. Outside the body, HIV is easily killed by common disinfectants like alcohol or peroxide.

Could the extreme mutability of HIV someday result in a contagious virus, like, say, polio or measles? While theoretically possible, the natural drift of evolution is toward an organism's lower virulence in a given population, not higher.

So let's end on that tiny note of optimism.

• influenza •

Influenza, "flu" for short, is a universal human experience, familiar to most people as just a really bad cold. Flu is dangerous to the extent that it can lead to pneumonia, especially for the elderly, the malnourished, or individuals stressed by chronic lung or heart problems. The viruses that cause flu are wondrously prone to mutate, making the manufacture of vaccines an annual crap shoot. Usually nothing to worry too much about, flu is nonetheless potentially capable of enormous havoc. Even in a normal year, twenty-five to fifty million Americans catch flu and from ten thousand to forty thousand die from it. When complicated by pneumonia, it is the sixth most common cause of death in the United States.

Influenza bugs are complex little machines by viral standards. Rugged spheres protected by layers of protein and fat, they contain long spiral chains of genetic material that look much like animal chromosomes. The spheres pirate a host by using more than seven hundred protein rods, called antigens, that fuse onto red blood cells and violate their membranes. In order to reproduce, the virus's chains must untangle, a process that offers many opportunities for "mistakes" to be made that might kill the offspring or make future generations more potent—or anything in between. The human immune system's antibodies are attracted to the rods, but the viral antigens may change or "drift" over time like a moving target. A

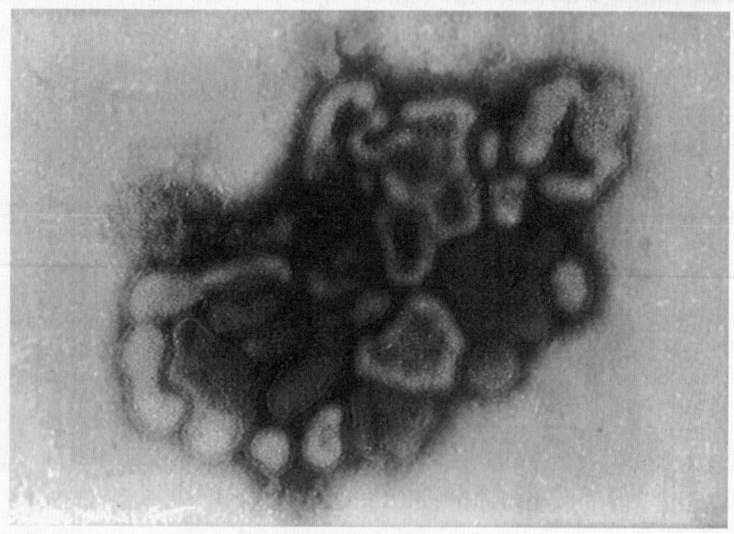

Gerald Ford's nemesis, the swine flu virus, properly called A/New Jersey/8/76 influenza.

successful bout against one variety of flu virus thus does not preclude a rematch with a slightly different opponent. If a new mutation of special strength pops up anywhere, resulting from a major change called an antigenic "shift," it will quickly spread around the world. All in all, an exasperating scenario for researchers, to say nothing of bedridden hosts.

Influenza epidemics are as old as human history, though they are hard to trace because of the generic nature of symptoms like fever, fatigue, coughing, and achiness. (Contrary to popular belief, influenza viruses do not cause intestinal problems; "stomach flu" is thus a misnomer.) The mysterious deadly plague of Athens in 430 B.C. may have been partly due to highly virulent flu. Charlemagne's army may have been decimated by the disease during an epidemic in 876. There is unmistakable evidence of deadly influenza outbreaks during the sixteenth century, with the first recorded pandemic occurring in 1580 across Africa and Europe.

By 1647, influenza was reported in North America as an epidemic moved up from the Caribbean into New England, killing thousands. The disease was known by colloquial names ("la grippe," "jolly rant," "the new acquaintance") until after the epidemic of 1732–33 in the American colonies, when an English doctor named John Huxham introduced an old Italian folk term that connected colds, coughs, and fevers to the astrological "influence" of the stars. Though it was a common killer generation after generation, flu did not attract serious medical attention during the centuries when other diseases were far more dramatic.

This situation changed with the great pandemic of "Spanish flu" in 1918–19, the twentieth century's worst of any kind and the deadliest faced by modern Western society. Spain was not fond of this term, needless to say, especially since the epidemic apparently started in the United States. Conservative estimates of worldwide deaths stand at about 21 million, out of perhaps a billion cases. India lost at least 12 million. The reliable figure for fatalities in the United States is 550,000. One tenth of American workers was sick in bed during the winter of 1918. One fifth of the U.S. Army and 28 percent of civilians caught the flu. More than 20,000 New Yorkers were killed in the fall of 1918. The virus was not satisfied with urban dwellers: Inuit villages in Alaska were wiped out and Samoa lost 20 percent of its inhabitants. Physicians had little sense of why such a disaster was happening or what to tell people to do to avoid it. Though an influenza virus had first been isolated from chickens in 1901, it was not recognized as such until 1955. There have been periodic waves of flu since 1919—Hong Kong flu killed about seventy thousand Americans in 1968–69—but we are probably still in an overall trough following the big one.

Not surprisingly, this disaster triggered sundry research on the part of a no-longer-complacent scientific community. In 1931 an influenza virus was isolated from runny pig snouts, spawning the name "swine flu" and lending credence to the belief of some veterinarians that pigs had caught the flu from humans in 1918–19. The pig bug was, in fact, closely related to the human pest, but no one will ever know exactly what caused the great pandemic, since no blood samples were saved. So-called influenza A virus, now the most common, was found in the throats of infected people in 1933, influenza B in 1940, and the relatively unusual influenza C in 1947. These are the three basic flu germs, variants of which are popularly designated according to where they first strike—Hong Kong (B), New Jersey (A), Bangkok (A), etc. Since 1918, influenza A has undergone two antigenic shifts, in 1957 and 1968. Nature's kitchen for new flu viruses is probably a duplex of pigs and water fowl—yes, ducks and seagulls—that periodically pass them around to other creatures, including humans, in their guano. In 1980 an influenza A strain produced an epidemic in seals and caused conjunctivitis in humans who tried to help them.

Since there is no reliable way to avoid influenza and no drug to make it go away once you catch it, vaccine is the only medical defense. Every year, an educated guess is made about what kind of flu will predominate during the next winter season. For example, the U.S. Centers for Disease Control, after studying some 1,500 flu virus samples from 120 laboratories around the world, selected in March 1994 a vaccine made from three strains expected to prevail during 1994–95: A-Texas, B-Panama, and A-Shangdong (which replaced A-Beijing from 1993). Seventy million doses were then manufactured by four drug companies for those at risk of serious complications. Immunity to these strains developed within a week or two of vaccinees getting the shot. (Because the vaccine is made from killed viruses, you cannot get flu from a flu shot. Only people allergic to eggs should avoid flu vaccines, which are grown in chicken eggs.) For people over sixty, flu vaccine can reduce the disease's incidence by half. Vaccinees who catch flu despite the shot may experience milder symptoms. Only once has a massive attempt been launched to head off an epidemic through general vaccination, namely the 1976 fiasco under President Gerald Ford against swine flu. Ford told the American people that the 1918–19 virus was on its way back and signed a law providing $135 million for an immunization campaign that ultimately reached only about a quarter of the population. Along the way, the program attracted many prominent critics, including Ralph Nader and Albert Sabin, finally collapsing when cases of Guillain-Barré syndrome, a rare paralysis, appeared at ten times the normal rate among vaccinees. Happily, the flu epidemic never blossomed. Washington eventually paid some $93 million in damages, sending liability ripples through the courts that put a permanent crimp in the American pharmaceutical industry's enthusiasm for making *any* vaccines.

By the way, flu viruses last for hours in dried mucus.

• *Legionella* •

Some diseases are said to have hastened the decline of empires, but only one is famous for putting a hotel out of business. In July 1976 an American Legion convention in Philadelphia's venerable Bellevue-Stratford Hotel set the stage for the discovery of a new genus of pathogenic bacteria. Later named *Legionella*, the bugs had found lodging inside 221 people who had come for the revelries or simply frequented the hotel. Thirty-four died subsequently from acute pneumonia.

Understandably, the Bellevue-Stratford lost its pull and folded within a year.

Of course, *Legionella* was new only in the sense that humans had not been scientifically aware of it before. After five months of desperate detective work led to its identification, researchers took out old blood samples from previous mystery cases and verified that the same organism had caused similar outbreaks in a Washington, D.C., hospital in 1965 and a Pontiac, Michigan, health department office in 1968 (thereby proving, among other things, that germs are not afraid of medical facilities). They even thawed out a guinea pig from 1947 that had been injected with blood from a patient who died from something untraceable at the time, and found *Legionella*. "Legionnaires' disease" has probably been behind thousands of deaths from pneumonia every year in the United States for a long time, especially since air conditioning became part of urban life and perhaps since the unwashed masses accepted indoor plumbing. Between a thousand and thirteen hundred cases are reported annually, out of perhaps ten times more cases, with a fatality rate ranging between 5 and 30 percent.

The culprit in Philadelphia, *L. pneumophila,* is one of more than thirty *Legionella* species now known to exist around the world, at least nineteen of which cause pneumonia in humans. *L. pneumophila* and its brethren like to live in environments akin to puddle scum—wet, dark, and kind of rotty. At the decidedly un-rotty Bellevue-Stratford, air-conditioner cooling towers on the roof provided a conducive summertime abode, from which the germs circulated throughout the edifice in a fine infectious mist. But *Legionella* bacteria have also been found in shower heads and faucet pipes. *L. pneumophila* can survive in room temperature tap water for more than a year. Streams, ponds, and plain old mud may harbor them. Hot tubs, humidifiers, vegetable stand misters, decorative fountains—the list of possible habitats is unsettling. Chlorination may not remove the threat if the concentration of organisms is high.

Once a person becomes infected by inhaling enough bacteria, two courses of disease (properly known as legionellosis) may occur, depending on such factors as the victim's age, use of cigarettes or alcohol, and other conditions that are still unclear. Legionnaires' disease, the more serious, strikes 5 percent or fewer of people exposed to *Legionella,* takes two to ten days to develop, and kills up to a quarter of its victims if untreated. The comparatively mild Pontiac fever shows many of the same symptoms (high fever, chills, headache, cough), but does not include pneumonia, hits 95 percent of those exposed, comes on in just five to sixty-six hours, and is debilitating for a week or so but not fatal.

Neither disease spreads person to person—you need to have breathed the original mist of contaminated water, and quite a bit of it. Thanks to the unlucky Bellevue-Stratford, regulations concerning the maintenance of big air-conditioner systems are more strict today than in 1976.

• *Listeria monocytogenes* •

The bacterium *L. monocytogenes* can cause potentially fatal food poisoning that leads to meningitis. It is especially dangerous for pregnant women and infants. *Listeria* bacteria are found everywhere and inhabit the guts of humans, birds, spiders, livestock, and other animals. *L. monocytogenes* has the especially bothersome ability to thrive on refrigerated foods, and therefore is sometimes encountered in cold delicatessen goodies. In October 1994 the U.S. government issued a warning against eating potato salad manufactured by I&K Distributors in Ohio and sold in twelve states, owing to *L. monocytogenes* contamination. Antibiotics can control it.

• Marburg/Ebola •

Marburg and Ebola are two exotic filoviruses that can earn big bucks in Hollywood. They guarantee high fatality rates (25 percent for Marburg, up to 90 percent for Ebola), produce grotesque symptoms, lurk in darkest Africa under murky circumstances—in short, possess all the ingredients for a blockbuster. They have already triggered far more melodrama than disease, as a result of two famous incidents when imported laboratory monkeys carried the microbes into the rich white man's world.

One day in the summer of 1967, three employees at a vaccine factory in Marburg, Germany, complained of flu; by the next day their condition deteriorated into what became weeks of acute viral blood poisoning, bizarre rashes, severe diarrhea, bloody vomiting, and hemorrhages from every orifice. Also, their skin came off. In Germany and Yugoslavia the number of patients among pharmaceutical workers, the doctors who treated them, technicians, and household contacts climbed past thirty. Seven died by the end of the year, others suffered liver damage, impotence, and psychosis. Investigators discovered that all the cases were linked by exposure to a shipment of African green monkeys from Uganda, half of which had died en route. They found a large new rod-shaped virus in the monkeys' tissue that resembled a woolly bear caterpillar, one that could roll up tightly for

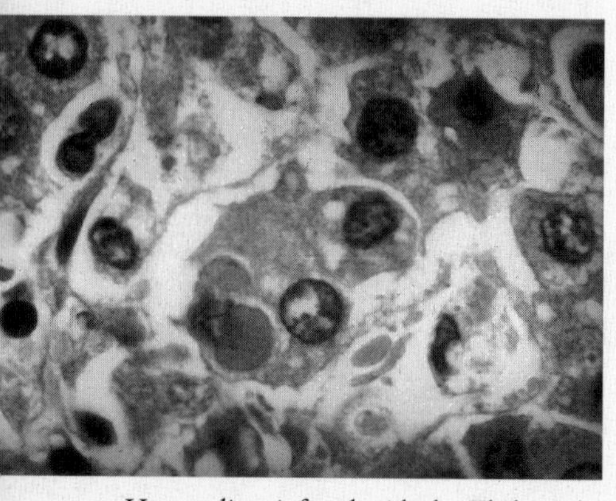

Human liver infected with the Ebola virus.

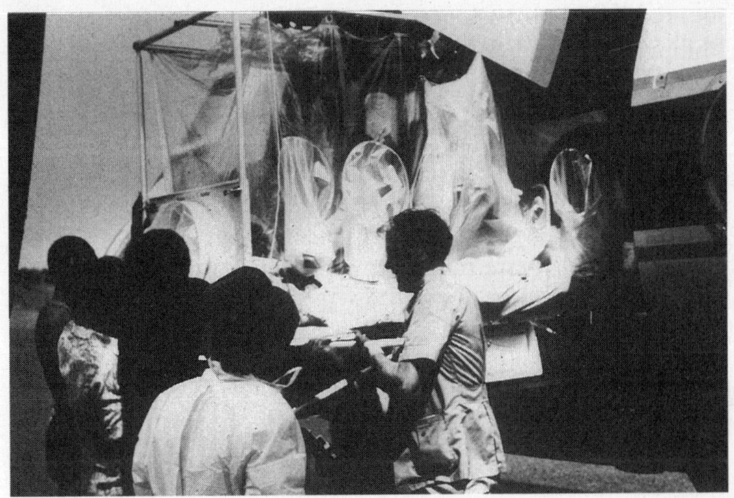

A Centers for Disease Control worker exposed to the Ebola virus being loaded onto an airplane while isolated in a plastic tent.

protection just like that cuddly little creature. No one has ever figured out exactly where the virus comes from, how humans catch it, or much of anything else that you would like to know about such a monster. Methyl alcohol knocks it out.

In the summer of 1976, panic swept remote villages in Zaire and Sudan in the wake of sickness and death in a form similar to the Marburg cases, only worse. Local physicians at first suspected yellow fever or typhoid, then Lassa fever, but soon called for outside help. Investigators discovered a virus shaped like a question mark, surely one of nature's all-time worst jokes. Named Ebola after a river near the epidemic's starting point, the virus turned out to be closely related to the Marburg germ and just as mysterious. Of 358 cases in Zaire, 325 died. In Sudan the virus was apparently somewhat weaker, killing 151 out of 284 victims.

In 1989 the Marburg plot recurred in the tony Washington

suburb of Reston, Virginia, when monkeys imported from the Philippines started dropping like flies at an animal quarantine facility. A virus similar to Ebola was soon isolated, causing feverish prayers among lab workers that the nation's capital would be spared, until it became evident that this new strain was not a people killer: several people were infected by the "Reston" virus, but none showed any symptoms. The scary episode further besmirched the reputation of imported monkeys and made a small fortune for at least one magazine writer.

In addition to underlining the necessity for strict international controls on animal trafficking and the potential for vicious diseases to escape from remote areas through fast modern transportation, the Marburg and Ebola outbreaks demonstrated one of the basic tenets of infectious disease: if a germ is exceedingly deadly, it will roar through a population quickly rather than cause massive, prolonged epidemics. Germs need fresh, healthy hosts, not corpses.

• measles •

Measles is *the* single most infectious common disease—one person with measles who blunders into a crowded room will give it to almost everyone. If there were no immunization, measles would be universal. Epidemics used to happen like clockwork every two to three years. That the United States experienced large outbreaks in the late 1980s and early 1990s among inner-city African-American and Hispanic children is shameful evidence of the weakness of basic clinical and laboratory services.

Measles is caused by one of the paramyxovirus gang, others of which bring mumps and sundry respiratory illnesses. The virus travels from victim to victim primarily in fine droplets sprayed by coughing or sneezing. About two weeks after exposure, the well-known rash appears with high fever. New viruses will grow in cells throughout the body, but the mucous linings of the respiratory tract are their best breeding ground, providing a perfect launchpad. Particularly in Africa, measles causes deaths from pneumonia, diarrhea, and malnutrition. People who recover are immune for the rest of their lives.

A thousand years ago, the disease was so ubiquitous among children that its earliest reporter, the Persian physician Abu Bakr Muhammad ibn Zakariyā, known as al-Rhazes of Baghdad (850–923), believed it was simply a natural episode of childhood—like losing baby teeth—rather than an infectious illness. The measles virus is not carried by any animal other than humans, and it needs a local population of between 300,000 and 400,000 to set up a sustaining, continuous supply of virgin bodies. Otherwise it dies out. Such settlements did not exist before the Sumerian cities in the Middle East, around 2500 B.C. Thus, medical historians theorize that measles evolved sometime after this date from a virus in cattle or dogs. Neither Greek nor Roman writers made note of it, but it might have been classed together with other severe fever-rash afflictions. Al-Rhazes was the first to distinguish between measles—which he called *hasbah,* meaning "eruption"—and smallpox, though he thought they had the same cause, a view that persisted for nearly eight hundred years. The word "measles" did not appear in English until the fourteenth cen-

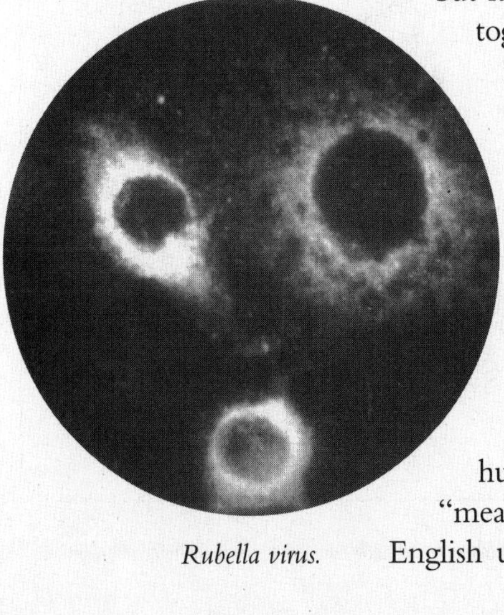

Rubella virus.

A Japanese woodcut, circa 1900, showing measles as an especially hirsute demon.

tury, stemming from the word "miser," which was used to refer to the wretchedness of lepers.

The disastrous effect of this germ on previously unexposed populations was most spectacular during the early colonial era. It was instrumental in cutting the population of central Mexico from about 30 million to 3 million within fifty years of Cortés's arrival. In otherwise healthy, well-nourished children, measles is not usually a killer. Yet 800 children died in the Charlestown neighborhood of Boston in 1772. In 1848, measles struck Hawaii for the first time, apparently imported from California, and killed 27 percent of some 150,000 cases. If these figures sound incredible, one need only consider the arrival of the H.M.S. *Dido* in the Fiji Islands in 1873. Within three months, a quarter of Fijians, some 30,000 people, were dead from measles. The virus is known for remarkably consistent potency: in 1951, measles commenced with a single case among 4,300 never-before-infected Greenlanders, and within six weeks all but five of them caught it (the mortality rate was only 1.8 percent, thanks to proper care).

Measles vaccine was not approved in the United States until 1963, when a half million cases were expected annually. No doubt because of the disease's close historical association with smallpox, a technique had been attempted as far back as 1758 that was similar to Edward Jenner's inoculation of healthy individuals with pus: blood from measles patients was smeared on scratched arm skin. This crude measure was not as successful as the equally primitive smallpox prevention with pus, however. Researchers were unable even to grow the virus outside the human body until the late 1930s. By the mid-1950s, they had learned how to measure how much virus was present when cultivated in chick embryos, thus opening the way for production of a safe vaccine.

Unlike the smallpox virus, which has been eradicated in

nature, the far more infectious measles virus still lurks everywhere ready to strike the nonimmune. In 1977 the United States set a goal of conquering measles by 1982. Reported cases fell to fewer than 3,000 a year, but then shot up to 25,000 in 1990. In the spring of 1994, an infected skier in Colorado passed the disease to hundreds of unvaccinated students, many of whom were Christian Scientists, who shun medicine in favor of prayer.

All children should be vaccinated at fifteen months (a bit younger in underdeveloped countries) and get booster shots later, but inadequate government support for public health has made this unachievable in many countries. In the United States, only 70 percent of two-year-olds were immunized in 1990. Because immunization programs must be nearly perfect to counter the virus's extremely efficient reproduction, war and lesser evils can easily derail even the best intentions. About a million people, mostly children, die every year around the world from measles.

A disease similar to measles but comparatively benign and less infectious is caused by the unrelated rubella virus. Also known as "German measles" because of research done by German scientists, it would be of little consequence if not for the fact that unborn babies are at high risk for birth defects if their mothers contract the disease during the first trimester (at least 20–25 percent risk of birth defects). This was discovered only in 1941 by an Australian ophthalmologist, Norman Gregg, who reported a link between mothers who had caught rubella and infants with congenital cataracts. After centuries of considering rubella to be trivial, the medical establishment balked for years at accepting Gregg's hypothesis, even attacking him editorially in the prestigious British medical journal *The Lancet* in 1944. But further research proved that rubella could indeed cause a bevy of defects, including heart problems, deafness, language disorders, and bone lesions.

Large epidemics continued to occur regularly around the world until a vaccine became available. Production of a vaccine was first licensed in the United States in 1969, when about fifty thousand cases were occurring every year, and has since reduced the disease to negligible importance. As long as kids get their MMR (measles, mumps, rubella) shots, the reservoir of germs that might infect women of childbearing age will remain nearly dry.

• mumps •

Mumps is the disease with a name that sounds just like the way it makes you feel. Hippocrates described an epidemic of mumps, well known across the ages for painful swelling of the parotid (salivary) glands in children and, in about a fifth of older male sufferers, of the testicles. Like measles, mumps is caused by a paramyxovirus (*see* page 90) transported into the respiratory tract via fine droplets from an infected person, but is far less dangerous and contagious than its cousin, being similar to flu or rubella.

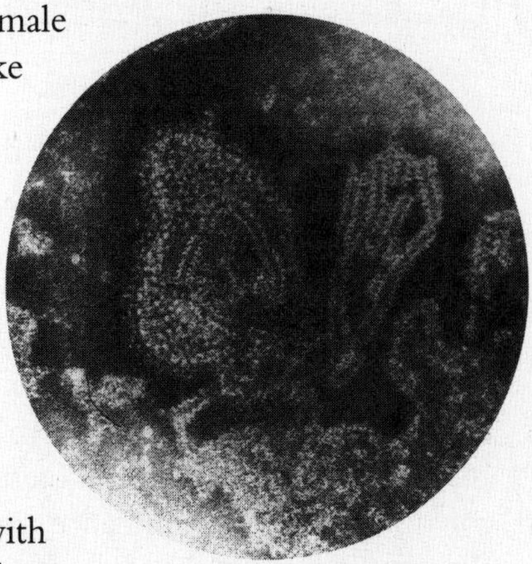

Mumps virus.

Mumps comes and goes with no apparent symptoms in about a third of all cases. Contrary to popular fear, it rarely causes sterility in adult men, though many of them unlucky enough to catch it must swear that's going to happen. It hurts. Because the virus likes glandular and nerve tissue, mumps can be accompanied by complications in many

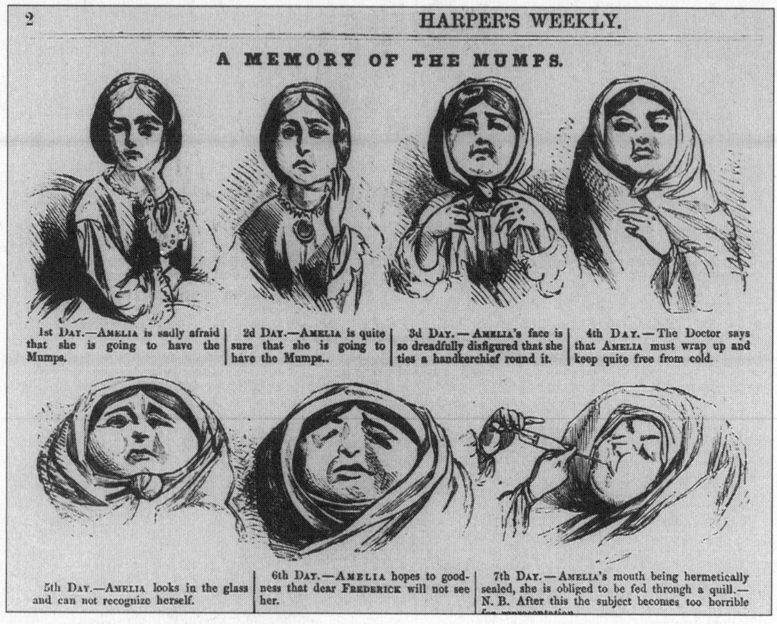

2 HARPER'S WEEKLY.

A MEMORY OF THE MUMPS.

1st Day.—Amelia is sadly afraid that she is going to have the Mumps.

2d Day.—Amelia is quite sure that she is going to have the Mumps..

3d Day.—Amelia's face is so dreadfully disfigured that she ties a handkerchief round it.

4th Day.—The Doctor says that Amelia must wrap up and keep quite free from cold.

5th Day.—Amelia looks in the glass and can not recognize herself.

6th Day.—Amelia hopes to goodness that dear Frederick will not see her.

7th Day.—Amelia's mouth being hermetically sealed, she is obliged to be fed through a quill.—N. B. After this the subject becomes too horrible

organs—pancreas, brain, heart, thyroid—but they are seldom serious.

Before vaccine became available in the 1960s, mumps tended to be an urban affliction with fairly irregular occurrence. It was thus bothersome to the military whenever large numbers of men were mustered together from different locales—many had already been exposed and were therefore immune, but many were susceptible. While not capable of causing significant fatalities, mumps could be a dangerous nuisance for an army in the field. During World War I, some 231,490 cases in the U.S. Army resulted in 3,884,147 man-days of lost duty. Mumps ranked third, behind flu and gonorrhea, for putting otherwise healthy soldiers out of action. There were even secondary attacks, especially among buglers. Discovery of the mumps virus in 1934 brought better understanding of how it was transmitted. The Army had only

103,055 cases during World War II in a vastly larger force, probably because the general population of recruits was less rural and therefore had been more evenly exposed before being drafted.

Mumps vaccine is usually given to one-year-olds in combination with measles and rubella vaccine (MMR). The total of 1,692 cases reported in 1993 was the lowest number ever in the United States. A worldwide immunization campaign could relegate mumps to archival status.

• mycobacteria •

Two organisms of the genus *Mycobacterium* are infamous for causing tuberculosis and leprosy, but there are myriad other types. They live harmlessly in soil, water, insects, synergistically around plant roots, and in many animals besides humans. Infections in people are probably due to contact with dirt, dust, or water. In children, old people, and particularly the immune-impaired, they sometimes cause lymph-system problems, skin ulcers (following cuts received in pools, fish tanks, or similar places), and, rarely, bone, urinary tract, and nerve disorders. These last, more serious infections are AIDS alarms.

• *Neisseria gonorrhoeae* •
(gonorrhea)

To the ancient Chinese, it was the disease that is "different from all others." To the Greeks, it was *strangouria,* or urinary blockage. The Hebrews, who went on poetic record as being extremely wary of loose women and adventuresses (Proverbs 5), referred to it as "discharge" (that's the polite translation, at least). The French, ever graphic, called it *chaude pisse*—hot piss—as far back as the twelfth century. The worldly Dutch said *klapoor,* from which the English got "clap." It is gonorrhea (from the Greek words *gonos*, "seed," and *rheein,* "to flow"), one of the oldest and most familiar diseases of humankind. "Pissing razor blades" sums it up quite well.

In the dark spectrum of venereal diseases (from Venus, the goddess of love), gonorrhea has a reputation for being not so

bad. It can ruin a relationship, but it won't really hurt anybody. This misunderstanding is probably a male view, ascribable to the fact that women suffer gonorrhea's worst consequences. The picture is clouded, moreover, by three common suppositions: most infected men experience mild discomfort and then get better automatically, antibiotics are available when necessary, and most infected women are symptomless. All are true, but there are important qualifications.

More than 90 percent of the men and women who infect each other will have no symptoms at the time. Mathematically speaking, the odds that infected males will transmit the disease to their sexual partners are between 50 and 70 percent. Some

20 to 40 percent of women, once infected, develop pelvic inflammatory disease—a no-nonsense complication that can lead to infertility—or develop what physicians ominously call a "constellation" of other gynecological problems. Women are less likely to give the disease to their uninfected partners (the odds are 20 to 30 percent, still way better than any casino game). But some 80 percent of men, once they've caught the bug, will then get what the ancient sages warned about. (Lesbians rarely give gonorrhea to each other. Gay men do, and may suffer from infections of the rectum and throat.)

It should be added here that nonsexual transmission of gonorrhea is so unusual as to be freakish. The bacterium that causes this illness, *Neisseria gonorrhoeae* (named for Albert Neisser, the German bacteriologist who discovered it in 1879), is exquisitely well adapted to certain human tissue and to ways of finding fresh

supplies, hence its primordial parasitic relationship. Infants can pick it up in their eyes during birth from infected mothers, and it used to be a prominant cause of blindness in newborns, but routine application of silver nitrate solution to the eyes of every child at birth has largely overcome this tragedy. Prepubescent children with gonorrhea have almost certainly been sexually abused.

Until 1937, when sulfa drugs were found to be effective killers of *N. gonorrhoeae,* there was no cure. Penicillin soon joined the fray. Unfortunately, the pest is very quick at learning how to resist antibiotics. Sulfa drugs were used so freely during World War II that resistant microbes popped up everywhere. By 1976, after the mushrooming of venereal disease that accompanied the advent of the Pill and the sexual revolution, penicillin-resistant strains (known as PPNG, for *p*enicillinase-*p*roducing *N*eisseria *g*onorrhoeae) began to appear in the United States, imported largely from the Philippines by military personnel. Microbes soon popped up that were also resistant to tetracycline. Other drugs can still combat *N. gonorrhoeae,* but the trend is obvious. There is no natural immunity, nor any vaccine. Some 3 million Americans are estimated to be infected every year.

Yet another excellent reason to use condoms.

• *Neisseria meningitidis* •

In 1974–75, a vast epidemic of meningitis, a disease in which membranes surrounding the brain and spinal cord become painfully inflamed, blanketed Brazil and was especially cruel to its poorest citizens in the squalid barrios of São Paulo, Rio de Janeiro, and other cities. Many victims fell desperately ill in minutes, with stiff neck and fever leading to hemorrhages, coma, and death within a day. The military junta responded with a vaccination campaign that reached some 80 million people—

an unparalleled feat perhaps only possible in such an unwieldy society when ordered by generals—but not before *Neisseria meningitidis* bacteria had sickened more than 250,000, killed more than 11,000, and left a third of the survivors neurologically damaged.

Such paroxysmal epidemics of *N. meningitidis* are uncommon, but because the microbes spread easily in overcrowded living conditions, they tend to claim lots of victims once they get started. An epidemic in Nigeria in 1948–50 took 14,273 lives out of 92,964 reported cases. The attack rate in Chad reached an incredible 11,000 per 100,000 people. Without antibiotics, mortality sometimes topped 80 percent in children. Much like the polio virus, *N. meningitidis* produces far more "silent" carriers than symptomatic infections, a factor that helps fuel hysteria in populations already shaken by the precipitous, seemingly random appearance of gruesome cases. The bacteria have also caused extensive outbreaks during modern military mobilizations, as experienced by American recruits during both world wars, Korea, and Vietnam. The disease was of such concern to the British army during World War I that studies were done to determine the precise separation distance of bunks that would minimize contagion. But like all types of meningitis, this one always focuses on children.

Though epidemics probably occurred at roughly twenty-year intervals in the nineteenth century after meningococcal disease was first described in Switzerland in 1805, various types of *N. meningitidis* were not identified until the 1940s and '50s. Widespread use of antibiotics followed these studies, but resistant strains appeared by the 1960s. The Brazilian epi-demic was complicated by the fact that two types of *N. meningitidis,* known as A and C, rampaged simultaneously—A was new to Brazil and was treatable with antibiotics, while C had been around before and was now highly drug-resistant. In the modern age of jet travel, this could recur in any big city with dense, impoverished neighborhoods.

Like the vaccine used in Brazil, which came from France, current vaccines in the United States are effective against both A and C bacteria. Several clusters of C infection occurred in the United States in 1993, but about half of all severe incidents of meningococcal disease in America are caused by another type, dubbed B. The general population is not immunized because incidence of the disease has been very low since World War II. But military recruits still get shots, as do individuals with sickle-cell anemia, which predisposes them toward *N. meningitidis* infection. If the germ ever starts ravaging an American city, innoculees could feel lucky for perhaps the first time in their lives.

• papillomaviruses •

"Barley-corn, barley-corn, Injun meal shorts,/Spunk water, spunk water, swaller these warts." Thus did Huckleberry Finn try to do away with the most familiar result of papillomavirus infection, the common skin wart.

Comparatively unpoetic scientists had described the viruses and associated diseases by the middle of this century, but because of difficulties in culturing them in the lab, research did not blossom until the 1980s. Over sixty-seven types of papillomaviruses are now known, plus an important link has been discovered between infection with certain ones and cervical cancer, which causes more than four thousand deaths every year in the United States. Moreover, evidence is accumulating that sexually transmitted types somehow interact with herpes simplex-2 and HIV to increase the risk of cervical cancer. The picture is not clear, however, since 10 to 50 percent of healthy women carry papillomavirus.

This virus has an affinity for skin and mucous membranes, hence the Huckleberry Finn bumps and the less well known venereal warts that can appear on the genitalia and anus. They are transmitted by direct contact and take an average of two to three months to produce a visible wart. Skin warts can be easily removed by freezing them with liquid nitrogen, but other kinds require more finesse, shall we say. So, yes, use a condom.

• parainfluenza and respiratory syncytial viruses (RSVs) •

As any parent knows, a group of little kids at the playground will usually include gigglers, criers, and coughers. With luck and forebearance, crying and coughing fade away as giggling is joined by whispering and shouting. Fortunately, we are concerned here with coughing, because it is far easier to explain than the rest of the din.

Bronchitis, bronchiolitis, croup, and pneumonia—four mainstays of the pediatrician's trade—are brought primarily to young children by members of the paramyxoviridae family of viruses, others of which cause mumps and measles. These diseases are universal because the bugs are, too. They are highly infectious through personal contact and need invade our bodies no deeper than our noses or throats to start replicating like crazy in the mucus there. The only way to avoid them would be to stop breathing. They strike early in life because they are *out there*. Most people grow partially immune to them, which is why parents at the playground aren't hacking as much as the kids.

The parainfluenza viruses—there are four types of medical interest—cause lower respiratory diseases in kids and upper respiratory problems in adults (most grown-ups have some immunity, so the bugs are defeated before they can travel very deep). They were first isolated in the mid-1950s from children with croup. The two types most closely associated with croup tend to strike tots under two in the autumn of the year, with the greatest number of cases in populous areas coming at two-year intervals. Boys seem to get croup more often than girls. The type associated with bronchiolitis and pneumonia works all year round, mostly on babies under six months old with no preference for blue or pink. So far, attempts to develop vaccines, perhaps best administered to infants as a nasal aerosol, have not met with much success.

The respiratory syncytial virus (RSV) was first found in a chimp with a cold. It is now recognized as the prime agent of life-threatening diseases of the lower respiratory tract (pneumonia, bronchiolitis) in infants and children all over the world. It is so universal that all adults are believed to carry an antibody, as do babies for the first few weeks of life until the resistance acquired from mother wears off. RSV is the only virus that causes pediatric epidemics of respiratory disease every winter in big cities. As with so many other ills, poor kids are more likely to suffer from serious infections, perhaps because affluence somehow helps delay exposure past the period of greatest vulnerability (the first six months of life) to an age when reactions are milder. Efforts to make a vaccine have been about as futile as those against the parainfluenza virus. The matter is complicated by the fact that an infant's defensive system against RSV invaders is suppressed by RSV antibodies carried out of the womb, so that any vaccine would have to finesse this obstacle at least until those antibodies decayed.

• *Plasmodium* •
(malaria)

Worldwide, about 400 million people suffer every year from malaria, the disease caused by four species of this little parasitic animal. We can venture to use the word "animal" here since

And don't ever bother to come here again.

Plasmodium is a genus of protozoans, which are considered to be higher on the evolutionary ladder than viruses and bacteria. At one stage in plasmodia's curious reproductive cycle they even come in two sexes. Malaria—from the Italian *mal aria*, meaning "bad air"—is of primary concern throughout tropical Africa, where more than 80 percent of the world's cases and 95 percent of deaths occur, and in rural areas of Asia and Latin America. But it is not entirely absent elsewhere, including the United States, where there are about 1,200 cases every year, usually among immigrants and international travelers. Somewhere between one and three million people, mostly children, die from it annually—a staggering toll for a disease most of us probably think of as belonging in the distant past, one that even experts believed was headed for extinction not long ago. In many ways, malaria is stubborn.

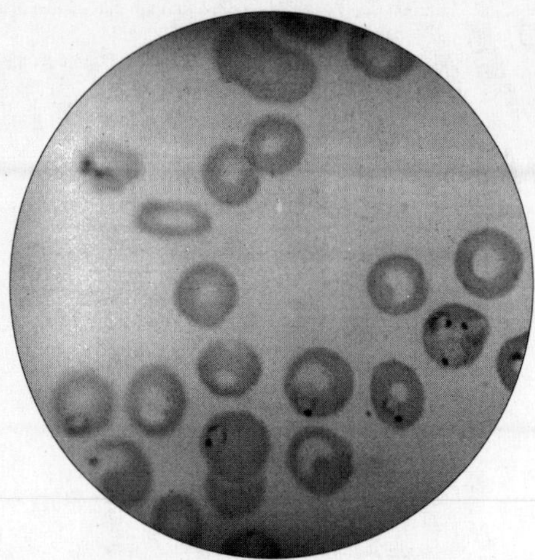

Red blood cells infected with plasmodium, an agent of malaria, magnified a thousand times.

Clear references to an affliction of intermittent fevers can be found at the start of recorded history in ancient Chinese and Sanskrit. Hippocrates described it in clinical detail; historians think it played a significant role in the cultural decline of Greece. The disease was so ferocious around Rome and along the west coast of Italy that some areas remained uninhabitable until the twentieth century. In Africa, it prevented European explorers from making contact with indigenous peoples as surely as any uncharted jungle or desert. More than 40 percent of British soldiers in India during the mid-nineteenth century suffered from malaria.

Since the 1890s, malaria has been understood to be spread not person to person but via mosquitoes, specifically certain female *Anopheles* species that feed on warm-blooded animals. The males are innocent vegetarians. (For ages, people were well aware that living around odoriferous swamps and low areas could bring on deadly "black-water fevers" or "agues," an empirical observation that was part of the "miasmatic" theory of disease.) While these spindly little Nimrods—who do a distinctive headstand with their hind legs in the air when they land—can be found throughout North America, they need a meal already carrying malarial plasmodia to get their chain of horrors going. Historical evidence supports the arrival of such microbes in the New World with Spanish explorers and African slaves, where they became a major problem for settlers in warm, wet environments like the Chesapeake Bay and Mississippi valley. George Washington and Andrew Jackson knew malaria personally, as did Napoleon. Indian tribes in these and many other regions were decimated by it. In the first decade of the nineteenth century the Illinois territory was likened to a graveyard because of malaria. The South was especially ravaged by the disease: divisions of the Union army fighting below the Mason-Dixon Line during the Civil War suffered 1,213,814 cases, including General Ulysses S. Grant.

Today the disease is rare in the United States and other developed nations, where drugs and insecticides (especially DDT after World War II) have largely obliterated domestic germs. Case rates in the United States fell from 34.7 per 100,000 population in 1946 to 1.4 in 1950. But travelers from malarial countries can still start local outbreaks, as long as lady *Anopheles* is around to help. In August 1993, three people in New York City were diagnosed with malaria apparently caught from local mosquitoes.

The cycle by which malaria germs thrive and spread is one of nature's darksome wonders. Whenever the right kind of mosquito bites a person with malaria, she sucks blood containing *Plasmodium* cells called gametocytes, which will reproduce sexually (there are male and female types) only inside this insect and nowhere else. The offspring, which are asexual and known as sporozoites, then get injected into any human unlucky enough to be the mosquito's next meal. They congregate in this victim's liver and start multiplying into a form called merozoites. An average of two to four weeks later, the merozoites begin attacking red blood cells, where they feed on hemoglobin and multiply into more merozoites plus a smaller number of new gametocytes. Unless another mosquito comes along at this point, these gametocytes will die and the whole process will reach a dead end—literally, because the destruction of red blood cells has released a toxin that brings on the paroxysmal chills and fevers of malaria; the most prevalent and deadly malarial bug, *P. falciparum,* may cause fevers as high as 110°F. Thus, although a reservoir of bad protozoans (in infected humans) is needed to get an outbreak going, epidemics won't happen without lots of mosquitoes. Note: Malaria can be transmitted by transfusion as long as those sporozoites stay in the bloodstream; one *Plasmodium* species' sporozoites can remain viable for more than forty years.

Obviously there is a chicken-egg problem here regarding how malaria came into being: Who was the first *Plasmodium* carrier, man or mosquito? How did the crazy system of 'zoites and 'cytes evolve? Never mind. By now the cycle is a great mandala of misery in many parts of the world, especially since both the mosquitoes and the protozoans have developed resistance to various insecticides and drugs. In the primordial West African breeding grounds of malaria, natives adapted eons ago to the threat with a gene that produces sickle-shaped, instead of round, red blood cells that are harder for malarial plasmodia to infest. Biologists calculate that the risk of death from malaria must have been at least 25 percent among people without the mutation. It was a mixed blessing, however: children who inherit this gene from both parents tend to die young. About 0.3 percent of African-Americans suffer from sickle-cell anemia.

As early as 1630, the corregidor, or Spanish magistrate, of Loxa, Peru, was relieved of malarial fevers by drinking tea made from the bark of the cinchona (his wife was the countess

Chinchon) or "quinaquina" tree, a balm no doubt used by Amerindians for ages. This remedy was a monumental discovery, considering that the standard method of treatment among European doctors included bleeding the victim, whose red blood cells were already being annihilated. Jesuit priests spread knowledge of the bark's power to Europe, where it gained acceptance, at least among Catholics, after use by Louis XIV. In 1658 Oliver Cromwell chose death over use of "Jesuit powder" to cure his chronic malaria. "Much are we beholden to physicians who only prescribe the bark of the quinaquina, when they might oblige their patients to swallow the whole tree," remarked the English wit David Dalrymple in 1785.

The medically useful agent called quinine was extracted by French chemists in 1820 and became one of two active ingredients in innumerable gin-and-tonics quafed on sultry tropical evenings by white imperialists. With the establishment of Dutch cinchona plantations in Java in the 1850s, Europeans

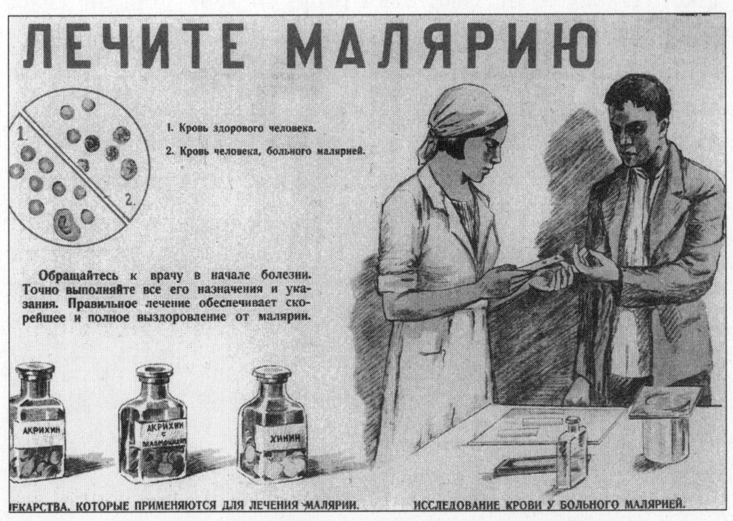

"The War with Malaria," a Soviet educational poster.

could proceed with their empire building in vast parts of Africa previously guarded by *Plasmodium* vectors. The Panama Canal was made possible by quinine and a massive antimosquito campaign in a zone that had been notorious for malaria. Some 500,000 cases of malaria occurred among U.S. forces during World War II, and after the Japanese conquered Java in 1942, synthetic drugs—derived from dyes developed in Germany between the world wars—were quickly developed and have replaced natural quinine.

The eradication of malaria from Sardinia in 1951 raised hopes for wiping it out everywhere. Between 1955 and 1965 the World Health Organization oversaw the expenditure of $1.4 billion to conquer the disease, but failed. In 1975, Europe was declared free of malaria for the first time in history, yet at that time the disease's worldwide incidence was about 2.5 times greater than it had been fifteen years earlier, having leapt from 1 million to 9 million cases in China and from 1 million to 6 million in India.

Today, malaria is becoming more dangerous as plasmodia grow resistant to the man-made drugs. In 1987 all of the Peace Corps workers in the West African nation of Benin came down with the disease while taking chloroquine, for four decades the standard penny-a-dose antimalarial medicine. An experimental vaccine called SPf66, which does not prevent infection but makes illness less likely, has been shown to be about 30 percent effective among village children in Tanzania, where malaria parasites hit almost everyone by age five. In many of the prime malarial regions of the world, the disease is considered out of control.

• pneumococcus •

Lots of microorganisms can cause pneumonia, but the "pneumococcus" is the chief creature behind lobar pneumonia, which damages an entire lobe of the lungs rather than small patches. Identifying it precisely was something of a Holy Grail mission for researchers for years, and more than a dozen scientific names were tried until *Streptococcus pneumoniae* was settled upon by the early 1970s. Pneumonia's popular monikers are far more revealing of its hideous potential than its Latin name: "captain of the men of death" and "friend of the aged."

Louis Pasteur first isolated the pneumococcus bacterium in 1881 from the saliva of a baby who had died of rabies. Investigators subsequently found more than eighty types of pneumococci as they tried to develop vaccines (gold and diamond miners in South Africa, who were dying like cattle under horrendous working conditions, were early recipients of only partly effective immunizations). Laborers on the Panama Canal, especially blacks, fell in great numbers to the pneumococcus. During World War I, pneumonia, fueled by flu and measles but nonetheless due most often to the pneumococcus, was the top cause of death in military camps. Modern vaccines now prevent widespread outbreaks in high-risk occupations like mining or the military.

Today the primary concern regarding *S. pneumoniae* is drug resistance, a disastrous phenomenon that could take us back to the days of Pasteur in a hurry. The prevalence of pneumococcal strains highly resistant to penicillin increased by a factor of sixty between the late 1970s and early 1990s.

• polio •

Perhaps no other disease illustrates the parochialism of the developed world more than polio. Since the introduction of the injected Salk vaccine, made from killed viruses, in 1956 and oral Sabin vaccine, made from attenuated live viruses, in 1960, paralytic poliomyelitis, which means inflammation of

ACUTE ANTERIOR **POLIOMYELITIS**

(A COMMUNICABLE DISEASE)

Keep Out of this House

By Order of BOARD OF HEALTH

HEALTH OFFICER

...erson removing this card without authority is liable to prosecution.

Board of Health quarantine notice, San Francisco, circa 1910.

the spinal cord, has become a relic of the past wherever immunization of children is routine. Certainly few Americans under the age of thirty-five have ever been haunted by images of iron lungs, steel braces, or aluminum crutches, let alone carried the disease on their personal worry list. Yet about a hundred thousand cases and ten thousand deaths still occur every year around the world, mostly in Asia and Africa, where immunization is anything but routine.

There are three distinct types of polioviruses, of varying potence, and they are the most dangerous members of the enterovirus group. They are bullets of Zenlike simplicity, consisting of little more than a single strand of twelve genes, targeted at their host's nerve cells. Polioviruses have been around since ancient times but did not gain much attention until an epidemic hit the United States during the summer of 1916. Though they never caused the massive fatalities of smallpox or yellow fever, the mystery of how they passed from victim to victim and the suddenness of their crippling attack on robust people spawned a level of superstition and phobia in otherwise "advanced" societies unequaled in this century.

Like other enteroviruses, polioviruses like to inhabit the alimentary canal. But not just any alimentary canal will do: only humans, monkeys, and chimpanzees are known to get polio—a fact that made vaccine research extra laborious, since scientists would much rather experiment with pliant little creatures like mice or guinea pigs. Polioviruses also affect far more people benignly (99 percent of all infections) than with obvious symptoms of disease. Until this fact was fully understood in the 1940s, one of the great puzzles of polio was why such a contagious germ caused relatively few incidents of serious illness—about 21,000 cases of paralytic polio occurred every year in the United States just before the Salk vaccine became available, out of a total population around 150 million. Vast numbers of hidden, but still communicable, infections also explained why quarantining known cases was an ineffective way to halt epidemics.

In the prevaccine era, polio scares were a phenomenon of the well-off industrialized world. As an enterovirus, polio is passed among people via feces from infected individuals, which means that people living in societies with excellent hygiene are more likely to grow up without having their first, naturally immunizing encounter with polioviruses at an early age when serious disease is a less likely consequence. Scientists even used to think that the poliovirus did not stalk the tropics, because paralysis was so rare in poor equatorial countries. Actually, infections did abound, but were asymptomatic, with deaths from "infantile paralysis" getting lost among already high child mortality rates. The prime age for polio in the United States by the 1950s was five to nine years old, with two thirds of deaths in victims over fifteen. Polio became notorious for striking the "best" families, Franklin D. Roosevelt being the most illustrious example. It was a dangerous side effect, so to speak, of affluence.

Killed-virus vaccine like that pioneered by Jonas Salk has not been manufactured in the United States since 1965. We now rely on the cheaper (pennies a dose), more effective live vaccine for universal immunizations—the number of annual cases of nonimported paralytic polio in America fell to four in 1985. Risk of domestic exposure to "wild" polioviruses (as opposed to the tamed viruses in vaccine) is now so negligible that adults are not immunized. Because live vaccine contains a weakened type of poliovirus that could in theory mutate into more virulent forms—an exceedingly rare, but not unheard-of, catastrophe—it is not given to people with impaired immune systems or to members of their households. Vaccine viruses are rigorously tested by injection into monkeys' spines.

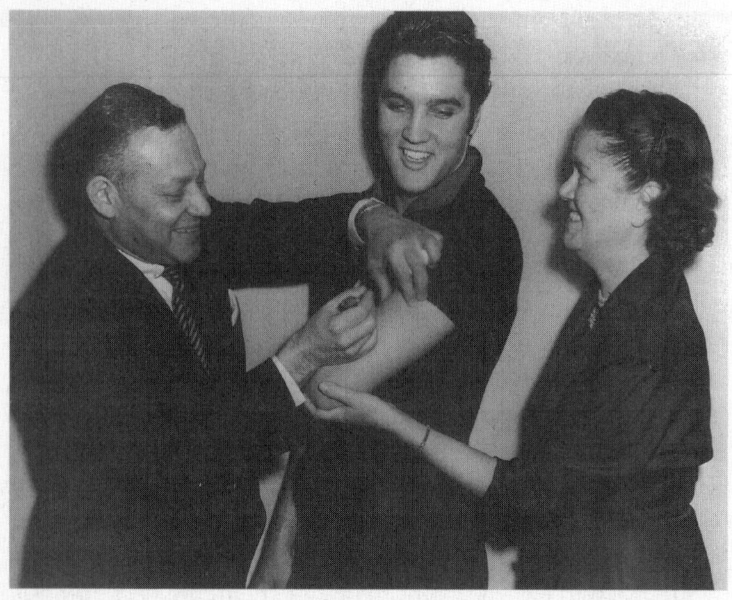

Looking not at all shook up, Elvis gets his polio shot.

Since 1988 the World Health Organization has led an effort to wipe out polio by the end of the century. In the Western Hemisphere the program has been successful. By the end of 1991, wild poliovirus had been found in only eight people (all children) living on this half of the planet, with person-to-person transmission of the germ apparently confined to Peru. Since then, the number has hit zero. As for the rest of Earth, 42 percent of all reported cases in 1993 occurred in India and 19 percent in Pakistan. Africa accounted for another 15 percent. Until polio is totally eradicated, countries deemed disease-free will be at risk from viruses imported from such regions.

• Q fever •

This generally nonlethal disease stumped investigators when first studied among Australian slaughterhouse workers in the 1930s; the label "Q" stands for "query." It is now well understood to be caused by a rickettsialike microbe, originally named *Rickettsia burnetii,* after F. M. Burnet, the Australian who first isolated it in the late '30s, but now called *Coxiella burnetii,* after H. R. Cox, an American who did similar work at the same time. It is transmitted to humans by livestock or, rarely, ticks, and is perhaps best known among pathologists for having more nonhuman reservoirs than any other human illness. Dozens upon dozens of different ticks, flies, lice, chiggers, birds, and mammals may carry the germ for Q fever. It just doesn't bother us much—a lucky thing, since only a single organism is needed to cause a full-blown infection.

Q fever's symptoms are much like those of typhus or Rocky Mountain spotted fever, only milder and with no rash. Just a handful of cases turn up every year in the United States and northern Europe, but sporadic outbreaks can occur wherever there are lots of cattle and sheep, especially among workers regularly exposed to livestock herds or products like meat carcasses, wool, and hides. *Coxiella burnetii* can survive for a long time in dust, making barns and other indoor livestock areas especially likely to harbor Q fever; in a classic outbreak during

World War II, hundreds of American soldiers camping on a farm near Florence, Italy, caught Q fever in a barn where movies were shown. Since it doesn't hurt the animals, nothing much is done about it besides pasteurization of milk, which has many other justifications.

Though not a mystery, Q fever still deserves its name because of one remarkable issue: How can an organism capable of massively infecting people from a minuscule initial exposure *not* cause serious problems?

• rabies •

Rabies has never been a common disease, but it has affected our metaphorical consciousness like few others. Shakespeare compared jealousy to "poisons more deadly than a mad dog's tooth." Sports fans and other raging fanatics are said to be "rabid." Extremely angry people are described as "foaming at the mouth," though this is usually a slight exaggeration. Shouting "wild dog!" has little dramatic value nowadays, but is still a good attention getter. Rabies is an animal disease, so unless we get in the way of sick warm-blooded mammals that decide to bite us, we will never, ever catch it.

Fascination with rabies is certainly nothing new. Mesopotamian codes before 2000 B.C. mention fatal dog bites. The Greek atomist philosopher Democritus (circa 420 B.C.), who traveled widely in Persia and Egypt, wrote about rabies in domestic animals, as did Aristotle a hundred years later. Epidemics (or, properly speaking, epizootics) of wolf rabies are known to have occurred in western Europe in the late thirteenth century. The first outbreak of dog rabies to leave an extensive record happened in Italy in 1708. The disease apparently did not exist in the Western Hemisphere until the mid-eighteenth century, when dogs caught it in Virginia in 1753. Louis Pasteur performed classic experiments with rabbits and dogs that led to the acquisition of an attenuated virus and the first human vaccination in a nine-year-old Alsacian boy named Joseph Meister, who thereby survived a rabid bite.

A cartoon published in 1826 shows a rabid dog causing panic in the streets of London—for good reason.

Though rabies has been eliminated in many countries through pet control and careful quarantines, it is endemic in various wild animals in the Americas and in domestic and wild animals throughout Africa and Asia, where most human cases occur today. About thirty thousand deaths are attributed to rabies every year, the vast majority in poor countries. Since 1980, twenty cases have been reported in the United States, half of which were imported.

The virus that causes rabies is said to be "neurotropic," that is, it seeks out nerve cells. Evolution has pumped such ferocity into this organism's craving for nerve tissue that it literally drives its victims crazy in order to jump to the next host, via the saliva in the rabid animal's bite wounds that break the skin in fact, the salivary glands actually become infected before most animals show any overt sign of rabies. The disease has a long incubation period, averaging a month or two in humans but sometimes years, depending a lot on the location (head, face, or hands are most dangerous) and severity of infective bites. This provides a window for treatment, which is generally futile after symptoms begin. Professionals are fond of saying that early diagnosis does not alter the patient's prognosis. Rabies victims at first feel a general malaise and restlessness, then grow increasingly agitated with painful spasms of the throat. Some start to hallucinate. Soon they cannot drink, which is why rabies has been called "hydrophobia." Death comes within ten days after the appearance of symptoms, though modern care can make this somewhat less than inevitable. Dogs actually show two forms of the disease: "furious rabies," the classic state of incoherent viciousness; and "dumb rabies," where paralysis takes over. Wild animals can get "furious," too, but they more often start by showing uncustomary behavior: nocturnal creatures come out in the daytime, the timid grow bold, etc. The last nonimported human case in California was due to a bobcat bite in 1969. An outbreak of canine rabies in South Texas has produced more than five hundred cases among wild coyotes and pet dogs since 1988, prompting a $2 million state program to airdrop 850,000 vaccine-laced dog biscuits over a 14,400-square-mile area. Two human deaths have been reported.

In the United States and Europe, where rabies in dogs and cats is largely controlled through vaccination, wild animals such as raccoons, skunks, bats, foxes, and coyotes are the virus's main reservoirs (humans are a dead end, since they do not transmit the virus and suffer essentially 100 percent fatalities if untreated). Of 8,644 animals reported with rabies in 1992, 43 percent were raccoons. Seven percent were bats, but bats are considered too important ecologically to exterminate like mosquitoes. Pets should not be presumed to be rabies-free. In October 1994, five hundred people in New Hampshire had to be immunized after contact with possibly rabid kittens from a Concord pet shop. The kittens, of which one died and tested positive and twelve others were destroyed and tested negative, may have been infected by a raccoon before delivery to the store. Rabbits and rodents, including squirrels, are almost never infected. Rabies can also be transmitted by means of infected meat, as Darwin observed among Peruvians who ate beef from a rabid bull. The virus may also be airborne, as a few unfortunate explorers of bat caves have discovered.

Most people bitten by a rabid animal will not catch the disease, but if a suspect critter is involved, vaccination should start as soon as possible anyway. There is only one method for testing an animal: kill it and examine its brain. Washing a bite thoroughly with soap is the best first aid. Though uncommon, the mere threat of rabies is expensive: although there have been just 20 confirmed American cases since 1980, 832 other people have required preventive vaccination, at an estimated cost of $850,000.

• rhinoviruses •

The rhinovirus is the king of colds, aptly named for the havoc it has been wreaking upon human noses since, well, since humans have had noses. Rhinoviruses come in so many different types, probably hundreds, that a universal cure is hopeless—you could spend a lifetime catching one new kind after another and never run out, which is exactly what happens. From Pliny the Younger, who in circa A.D. 100 prescribed "kissing the hairy muzzle of a mouse," to Sir Alexander

Fleming, who in 1945 recommended "a good gulp of hot whiskey at bedtime—it's not very scientific, but it works," cold remedies have run the gamut of imagination. But the rhinovirus endures.

Rhinoviruses just happen to thrive at the average temperature of our nasal mucous membranes, about 93°F. They are cousins of the poliovirus, possessing about half the same genes, which may be why they show the same tendency to strike in the late summer and fall (in the Northern Hemisphere, that is). No one has ever figured out exactly how they spread—close-up exposure to infected snot, pardon the colloquialism, is the best guess, accomplished by carrying germs on the hands and fingers up to the nose and eyes. Several enterprising microbiologists ran an experiment in 1973 that showed kissing ("direct oral contact") to be an inefficient way of spreading rhinoviruses. Other researchers found in 1977 that psychological stress makes rhinovirus colds worse. No one has taken the next leap forward to determine the effects of stressful kissing.

Though it is impossible to tell a rhinovirus cold from any other kind by symptoms alone, noses may get extra runny and coughs are especially common. Rhinoviruses are known to trigger asthma in children. A rhino cold lasts on average about a week, with peak misery on the second and third days. There is shaky evidence—far from conclusive—that heavy doses of vitamin C may reduce the length of colds, but not the frequency of catching them. Washing your hands a lot is a better bet: concentrated rhinoviruses in snot survive for hours on skin, plastic, wood, Formica, steel, and many fabrics.

So the old proverbs are as good as any other remedies. From England: Stuff a cold and starve a fever. From Germany: Sauerkraut is good. And from India, where they should know: One cold in the head is as bad as ten diseases. Keep in mind this useful rule of thumb: untreated colds last a week. Medical attention can end them after seven days.

• *Rickettsia* •

Creatures in the genus *Rickettsia* occupy a niche between bacteria and viruses. They carry much of their own cellular equipment and are vulnerable to antibiotics, but like viruses they need to invade living cells in order to grow. They were named after Howard T. Ricketts, an American scientist who identified them in 1907 and later died from typhus in Mexico.

Most rickettsial diseases fester among animals over the long term and reach humans via insects that carry the microbes. Rocky Mountain spotted fever is caused by *Rickettsia rickettsii,* which is transmitted by ticks. Typhus comes from *R. prowazekii* carried by body lice or rat fleas. Rickettsialpox, a mild fever-rash disease first seen in New York City, is carried by mouse mites infected with *R. akari.* Scrub typhus, found across the region of Japan, India, and Australia, comes from *R. tsutsugamushi* transmitted by chiggers from various rodents. Myriad types of tick-borne typhus are known simply by where they happen: Queensland, Marseilles, India, etc. In general, the Eastern Hemisphere's tick-borne diseases are milder than the Western Hemisphere's.

Because of the involvement of insects often associated with poor hygiene, rickettsial diseases used to mean serious problems for the military. Trench fever, a now-rare illness caused by *R. quintana* transmitted when infected louse feces are rubbed into itchy skin, got its name from the miserable conditions of World War I. Not much was known about it until the hellish confrontations in Flanders and France in 1915. Between 1915 and 1918 it accounted for one fifth to one third of all illness in British forces in France and for about one fifth in German and Austrian armies. By the time American soldiers arrived en masse on the Continent, the louse's role was understood, thanks to volunteers who allowed themselves to be fed upon by body lice from trench fever patients. The disease is still endemic in

Poland and Russia, as well as Mexico and northern Africa. Reports of its appearance among homeless alcoholics in the United States and France in the 1990s underline the erosion of public health care even in the richest societies.

Ehrlichia chaffeenis, a rickettsial bacterium first recognized in 1986, causes systemic infection with fever, headache, low white-blood-cell count, and sometimes meningitis in people bitten by a still undetermined tick. A germ almost identical to *Ehrlichia equi* is the cause of another recently recognized tick-borne illness called human granulocytic ehrlichiosis (HGE), which is spread by the same deer tick that can carry Lyme disease. It has emerged in Southern and Mid-Atlantic states. Nature never rests.

• Rocky Mountain spotted fever •

This is one of those uncommon diseases, like rabies, that have a way of generating more than their fair share of fear. Over the years, around five hundred to a thousand cases have turned up annually in the United States, with a fatality rate of about 7 percent in the most vulnerable patients if antibiotic treatment is started promptly. Yet many a parent will go ballistic at the sight of junior walking across a summer meadow, because *you might get a tick bite!* Kids should be examined for ticks after a day outside in infested areas, but really, Mom and Dad, let them play.

Rocky Mountain spotted fever is caused by *Rickettsia rickettsii,* an organism somewhere between viruses and bacteria on the evolutionary ladder (*see Rickettsia*). They are found naturally only in the Western Hemisphere (other rickettsia cause various tick fevers in the Eastern Hemisphere). Humans are accidental hosts and have no role in sustaining *R. rickettsii,* which depend on certain *Ixodidae,* or hard-backed, ticks—which include wood ticks, dog ticks, but *not* the deer ticks of Lyme disease fame—and the animals whose blood they suck. The name Rocky Mountain is somewhat misleading: though the disease was first recognized in those Western states, it has been encountered in all the lower forty-eight except Maine, with highest frequency in the Southeast.

The earliest reports of the affliction came from doctors in Idaho's Snake River region in the 1870s, who called it "black measles." Folk wisdom ascribed it to drinking melted snow or breathing sawdust. A physician in Great Falls, Montana, with the perfect name of Earl Strain, first perceived an association of the symptoms with tick bites in 1902. Then the illustrious young pathologist Howard T. Ricketts took up the subject at the University of Chicago between 1906 and 1909, describing scientifically the tick vector and the process of microbe transmission. In 1909, the Montana legislature voted $6,000 for his crucial research, but the state's board of examiners never sent him the money. Ricketts left for Mexico to study typhus in July of that year, where it killed him in 1910 at the age of thirty-nine. To honor his memory, the organisms that cause typhus and sundry spotted fevers were dubbed rickettsia in 1916.

Before the late 1940s, when antibiotics became available to cure most cases of Rocky Mountain spotted fever, mortality averaged about 20 percent (lower in children, higher in heavily exposed older adults). After an incubation period of three to twelve days, the victim gets a sudden, splitting headache with high fever, nausea, and chills. After about four days of fever a pink rash appears on the wrists and ankles that quickly spreads from the extremities to the trunk and gets darker. (Oddly enough, typhus rash spreads the other way, from trunk to extremities.) If untreated, severe infections lead to pneumonia, shock, and death. There is no vaccine licensed for general use. The disease is not contagious person to person, requiring an infected tick bite or exposure to tick feces (hence the warning not to crush a tick when removing it from the skin). *R. rickettsii* often require a meal of fresh blood to become activated into a virulent form, which means the tick must stay attached to a person for several hours in order for the disease to be transmitted.

Remember, a drop of whiskey helps disengage a hungry tick—more than enough reason to carry a flask in your pack.

• rotavirus/Norwalk •

About half of all babies and young kids hospitalized with diarrhea are infected with rotaviruses, which mess up the insides of many animals, too. Along with the so-called Norwalk (Ohio) viruses, rotaviruses play a monumental role in human health all over the planet, where there are billions of cases of diarrhea every year and millions of deaths in poor countries.

Both viruses were discovered in the early 1970s. Before then, the pathogens behind most cases of gastroenteritis and epidemic diarrhea in young children were a mystery; bacteria such as *Escherichia coli* are more prevalent in older children and adults. The Norwalk and rotaviruses pass around through infected feces. They are therefore classic "filth" agents, turning up in drinking water, swimming areas, uncooked foods, and along other fecal-oral routes wherever sanitation is awful. Vaccines are not available.

• *Salmonella* •

The leading cause of food poisoning, nontyphoid *Salmonella* bacteria have surged in medical importance during the past decade as a result of drug-resistant strains and lax food handling. Raw or undercooked eggs are a prime source, though only about one egg in every ten thousand carries *Salmonella*. Between 1985 and 1993, there were reports of 504 outbreaks that produced 18,195 cases, 1,978 hospitalizations, and 62 deaths. But far more Americans than that are infected every year—probably around one percent of the population, amounting to between 2 million and 4 million people. Even for many baby boomers who have never been sick, *Salmonella* holds a rancid place in modern history as the reason they had to get rid of their little green pet turtles in early childhood. Some of them may also remember when marijuana was rumored to be contaminated.

Salmonella poisoning is not especially dangerous as long as one's immune system is working properly or effective antibiotics are available for complicated cases. Gastroenteritis is the technical name for nausea and cramps that start twelve to forty-eight hours after ingestion of the bacteria, soon followed by diarrhea, fever, and maybe vomiting. The whole mess should go away in one to four days, leaving the victim adamant about never eating again whatever was at fault.

Salmonella is a very big genus, with 2,200 known members. (The surname, so to speak, comes from the American veterinarian Daniel E. Salmon, who first described the pests in 1885.) Some prefer humans, some don't, most are ubiquitous. The more common types are named after the place where they first spoiled a party: *S. heidelberg, S. newport, S. montevideo, S. saint paul.* One has an especially apt moniker: *S. agona.* The U.S. government established meat inspection laws in 1906 to prevent infections, but the record since then has been far from spotless. As new ways to get sick are discovered, new regulations are written; for example, milk and eggs that will be dried for myriad food products must first be pasteurized—a lesson learned the hard way, of course. In October 1994, thousands of Americans fell ill after eating contaminated ice cream from a Schwan's plant in Minnesota, leading quickly to new rules about when certain preliminary ingredients need to be pasteurized and the use of tanker trucks. The incident resulted in at least 489 confirmed cases in 29 states, with another 3,342 probable or suspect cases in 39 states. Any food that contains raw eggs is suspect—cookie batter, hollandaise sauce, Caesar salad, etc.

Multiple resistance to antibiotics has become a big bugaboo with *Salmonella,* especially since they can transfer their resistance to other gastrointestinal bacteria, such as *Escherichia coli.* Overabundant use of livestock and chicken feed laced with antibiotics is a primary force behind the evolution of drug-resistant strains that then find their way into people. Anybody already taking antibiotics for something rather ordinary, like an earache, is in for a bad ride if he eats a burger contaminated with *Salmonella,* since the very likely drug-resistant food bugs will land in territory that has been swept of competition.

So skip the beef—have a nice salad with oil and vinegar. At home, wash hands, utensils, and surfaces carefully with hot, soapy water after preparing raw chicken—more than one bird in three is probably infected.

• San Joaquin fever •

Inhaling the spores of the fungus *Coccidioides immitis* can cause the illness known as San Joaquin fever. Americans have to travel to dusty places in the Southwest to have any chance of catching it. Many infected people notice nothing, or maybe flulike symptoms at most. Some get a romantic-sounding but no doubt bummer affliction known as desert rheumatism. They'll all probably be okay with no treatment. But anyone with a weak immune system may—weeks, months, or years later—come down with an increasingly awful set of problems in his lymph nodes, bones, liver, brain, and other organs. Nearly two thirds of them will die. Progressive coccidioidomycosis, as this curse is called technically, is one of the definitive diseases for AIDS. An ongoing outbreak of "valley fever" in California started with

1,200 cases in 1991, increasing to 4,137 in 1993, mostly in Kern County. Long drought followed by heavy rain in central California may have helped spread *Coccidioides immitis*. Another cluster of cases occurred in Ventura County after the 1994 Northridge earthquake.

• smallpox •

Smallpox was the least understood and most destructive disease in history, yet the first to be eradicated from nature. It was caused by a fiercely infectious germ of the *Orthopoxvirus* genus, which includes a menagerie of other plagues: cowpox, rabbitpox, buffalopox, monkeypox, camelpox, raccoonpox, et cetera (but not chicken pox). One droplet of exhaled moisture from the lungs of an infected individual contained a thousand more smallpox viruses than the minimum necessary to produce the disease in someone else. The fact that we don't have to worry about this devil anymore shows what can be done with enough money.

Between its first recorded attack in ancient Egypt and the last in 1977, it claimed hundreds of millions of victims. Telltale signs of a smallpox rash on the mummy of Rameses V (who died 1156 B.C.) indicate the disease's antiquity. The Roman Empire, during a fifteen-year epidemic that began in A.D. 165, lost more than a third of its subjects in some areas. Crusaders brought smallpox back with them from the Holy Land—more than sufficient revenge by the infidels. Conquistadors carried it to the New World, where it mowed down the Aztec and Inca empires. The word "smallpox" began to be used in English during the sixteenth century as the translation of the French *la petite ve'role* (*ve'role = pox*); *la grosse ve'role* was syphilis. In the American colonies smallpox was a primary force allowing Europeans, who had developed immunities, to eradicate the culture of indigenous peoples, who had not. Pocahontas died

An illustration published in 1802 by a London antivaccination society shows Edward Jenner inoculating a young woman as former patients suffer the side effects of his vaccine, which consisted of fluid from a cowpox pustule.

of smallpox in 1617 after visiting London. In Europe it maintained the power to strike all levels of society, killing Queen Mary II of England in 1694, Emperor Joseph I of Austria in 1711, and King Louis XV of France in 1774. In Japan it claimed Emperor Gokwomyo in 1654 and Emperor Komei in 1867. George Washington survived a bout with the disease after visiting Barbados in 1751, but was severely scarred. The Boston Tea Party might have been a smallpox massacre had not the infected crew of the British brigantine *Beaver* been quarantined in December 1772 before joining other ships at Griffins Wharf. Isolated regions could be devastated, as when more than a third of the population of Iceland perished in 1707.

The quality of professional medical care during this era may be surmised from the record left by a smallpox patient of Thomas

Sydenham, one of the most illustrious English physicians: "In the beginning I lost twenty ounces of blood. He gave me a vomit, but I find by experience purging much better. I went abroad, by his direction, until I was blind, and then took to my bed. I had no fire allowed in my room, my windows were constantly open, my bedclothes were ordered to be layed no higher than my waist. He made me take twelve bottles of small beer, acidulated with spirit of vitriol, every twenty-four hours. I had of this anomalous kind to a very great degree, yet never lost my senses one moment."

In 1714 a Greek physician, Emanuel Timoni, published an article about how he had prevented the disease during an epidemic by embedding a knife in a victim's rash, then scratching a healthy person with the blade. This method, which must have seemed like pure magic, was apparently similar to age-old folk techniques known across China, India, and western Asia. In 1721 Lady Mary Wortley Montagu, the ebullient wife of the English ambassador to Turkey, stunned polite circles with the story of how she had permitted a Gypsy woman to immunize her infant son in a similar way. ("I am patriot enough to take pains to bring this useful invention into fashion in England," she had written to a friend in 1717, "and I should not fail to write some of our doctors very particularly if I knew any one of them that I thought had virtue enough to destroy such a considerable branch of their revenue for the good of mankind.") Children of the English royal family were then immunized after the king first tested the procedure on six condemned prisoners and eleven charity children.

That same year, Cotton Mather, who had heard of the practice from his slave, convinced fellow Congregational ministers to advocate such a desperate measure, as smallpox was threatening to depopulate Boston. Some 6,000 residents of the city contracted the disease and 900 of them died. Of 287 who braved primitive immunization, only 6 perished. Despite this success, inoculations were widely illegal in the American colonies, mainly because the often slapdash technique could itself start epidemics and was poorly explained to an understandably wary public (indeed, there was then no real explanation available). Benjamin Franklin, who greatly mistrusted the medical expertise of his day, could not rise above the counterintuitive proposition of exposing someone to a ghastly disease in order to save them from it—his four-year-old son died in a 1736 epidemic. John Quincy Adams, on the other hand, was inoculated at the age of eight in 1775. Educated Europeans also had trepidations: Mozart caught smallpox, which he survived, because his father believed preventive action interfered with God's will. But Catherine the Great of Russia, prodded by Voltaire, paid an English doctor £10,000 plus a lifetime annuity of £500—a fee that even modern physicians might envy—to inoculate her and members of the imperial court. Immanuel Kant thought the procedure was deadly, however. Only after George Washington ordered his army to line up and be inoculated during the Revolutionary War were anti-inoculation laws revoked.

Shitala Mata, Indian goddess of smallpox, sitting sidesaddle on a horse while holding a pot and a sheaf of fennel, perhaps on her way to a nasty business conference with Sopona.

In 1796 an English country doctor named Edward Jenner found a less threatening procedure that became one of the great milestones of preventive medicine. He had noticed that milkmaids who caught cowpox, a similar but relatively mild disease transmitted by cows, were somehow immune to smallpox. He therefore tried injecting an eight-year-old boy named James Phipps with fluid or "vaccinia" (hence the word "vaccine") from a cowpox pustule and, six weeks later, pus from a smallpox lesion. The brave boy reacted slightly to the cowpox inoculation, but did not catch smallpox at all. Though the Royal Society rejected Jenner's paper describing this groundbreaking experiment and twelve subsequent trials, he soon published a book that provoked widespread testing to confirm "vaccination" as safe. It was soon in general use and Parliament thanked Jenner with an award worth a then colossal $150,000. "You have erased from the calendar of human afflictions one of its greatest," Thomas Jefferson wrote to Jenner in 1806.

Sopona, god of smallpox—a wooden statue adorned with hair, shells, beads, and a sheaf of plants—worshiped by the Yorubas of Nigeria.

Still, smallpox remained a serious problem for more than a century outside the sphere of Jenner's vaccine, including the United States, where massive immigration and antivaccination societies kept health officials on the defensive. "A girl cannot be termed beautiful until she has had smallpox," pronounced a Sicilian proverb, reflecting the disease's ubiquity. Abraham Lincoln fell ill with smallpox several hours after delivering the Gettysburg Address. By the 1950s, compulsory vaccination of schoolchildren had largely freed North America, Central America, and Europe of the scourge. Hospitals were among the last major wellsprings of infection.

In other parts of the planet, however, there were still 10 million to 20 million cases of smallpox per year, causing 2 million deaths. The disease was so common in Bangladesh that the word for springtime, *bashunto,* was also the term for smallpox. In 1975 the last incidence of wild variola major, the most virulent form of the disease, was cured in a three-year-old Bangladeshi girl named

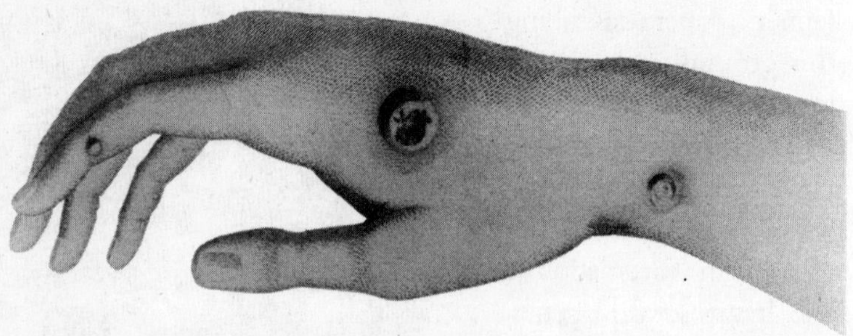

A drawing from "An inquiry into the causes and effects of the variolae vaccinae," published in London in 1798, showing the milkmaid's hand Jenner used as a source for his vaccine.

Rahima Banu. Finally, in 1977, after a $330 million, ten-year campaign (first proposed by the Soviet Union in 1958) supported by more than 150,000 workers in forty-one African, Asian, and South American countries, smallpox viruses were confined to research labs. The last natural case of smallpox, variola minor, occurred that year in a Somalian hospital worker named Ali Maow Maalin.

Today the last remaining germs are frozen with liquid nitrogen in about six hundred inch-long test tubes stored under maximum

The spirited Lady Mary Wortley Montagu (1689–1762), by all accounts magnificent in mind and body.

security at the Centers for Disease Control in Atlanta and the Institute for Viral Preparations in Moscow. These prisoners were slated for execution on New Year's Eve, 1993, but won a reprieve after some scientists insisted they could be valuable for future study of the human immune system. Knowing that the infamous *Orthopoxvirus* may become the first species to be made deliberately extinct, researchers drew a map of its DNA sequence for future reference. Death by pressure cooker at 248°F for two forty-five-minute cycles is still a matter of debate.

Fear that germs might accidentally escape—in 1978 a medical photographer died from smallpox after working one floor above a lab in Birmingham, England (the resident virologist later committed suicide)—has thus been put aside while the World Health Organization considers their fate. Careful surveillance for even a single new case of the disease continues, especially where infection might come from animals via mutated *Orthopoxvirus*. Russian scientists even dug up smallpox victims who had been buried in Siberian permafrost to see if the virus was present in their flesh; it was not. Military personnel are still vaccinated against potential use of the germ in bacteriological warfare; they occasionally catch smallpox from the vaccination and then infect others who come in contact with them. The United States alone stores seventeen million doses of vaccine, just in case.

• *Staphylococcus* •

In 1928, the British scientist Alexander Fleming found by accident that *Penicillium* mold could kill the dreaded *Staphylococcus* bacterium, thus launching the golden age of antibiotics. Fleming gave up on trying to isolate the specific substance involved, but scientists at Oxford succeeded in 1940 in isolating penicillin. During World War II, development of a usable product moved to the United States, where large-scale production and testing of penicillin took place—with the rather

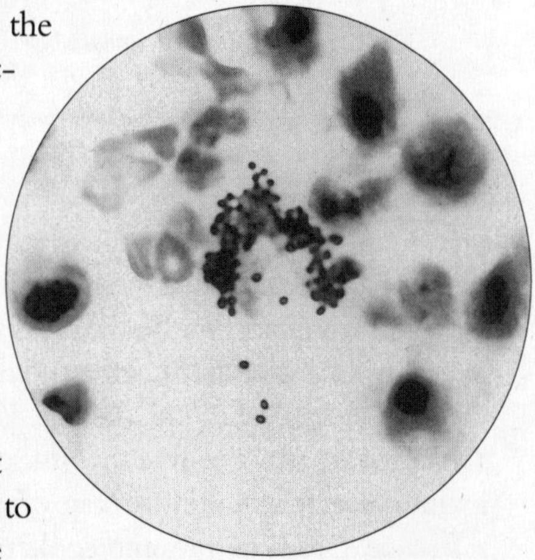

Staphylococcus *bacteria, bunched together like grapes, magnified a thousand times.*

unfair twist that the Brits had to pay royalties to manufacture the miracle drug they had discovered. Penicillin was truly a wonder: potent against pathogens in tiny amounts, safe to inject directly into the bloodstream, showing low toxicity even in high doses. Understandably, it became a universal lifesaver, but its very success led to an evolutionary backlash: the pathogens learned how to beat it. By 1982, penicillin could stop fewer than 10 percent of staph infections. There were alternative drugs, of course, but by the 1990s the widespread existence of superresistant *Staphylococcus* strains was a genuine crisis, especially in poor countries where the few remaining effective antibiotics are impossibly expensive.

Staph infections are no doubt primordial. The boils that Job scraped with broken pottery could only have been caused by staph. But the bacteria were not observed and studied until the late nineteenth century, when Sir Alexander Ogston of the Royal Aberdeen Infirmary named the organisms after the Greek words for grape (*staphyle*) and berry (*kokkos*). They grow like bunches of grapes everywhere. Thirty to 50 percent of healthy adults carry them harmlessly in their noses, and 20 percent have them on their skin. There are two species, *S. aureus* and *S. epidermidis* (originally called *S. albus* to differentiate its white colonies from *aureus*'s golden orange colonies). Most infections are confined to the skin and are not serious, let alone fatal. But if they spread and enter the bloodstream, the consequences can be dire. They are particularly prevalent in hospitals: of about 23 million surgeries in the United States in 1992, there were 920,000 that resulted in bacterial infections, most of which were staph, a rate rivaled only by *Escherichia coli*.

S. aureus, the more virulent of the two, produces a powerful toxin that can cause food poisoning, which is felt very quickly, in as few as thirty minutes after eating, compared to other stomach turners that take hours to kick in. Chinese mushrooms were taken off the U.S. market in 1989 when it was discovered that transporting them in plastic bags could help cultivate *S. aureus.* The microbe is highly contagious through contact with its colonies in wounds, and can bring on pneumonia. It is also the devil behind toxic shock syndrome—diagnosed in women who used superabsorbent tampons—which came to the public's attention in the early 1980s.

The human immune system is very good at combating most staph infections, so *Staphylococcus* bacteria prey hard on the weak: infants, the elderly, the malnourished, drug abusers (especially heroin addicts), the immune-impaired, the obese. Burn victims and flu sufferers are known to be more susceptible, as is anyone with foreign objects in his or her body (sutures, artificial valves or joints, intravenous tubes, tampons, etc.). But because of their ubiquity, *Staphylococcus* germs can strike anyone, given half a chance. George Bush experienced a severe staph infection in his right arm as a teenager, though it did not prevent his raising that arm to take the oath of office later in life.

• *Streptococcus* •

"Getting your tonsils out" used to be one of childhood's rites of passage—for many people the first and, with luck, the only stay in the hospital until old age. Plus you could have all the Popsicles you wanted to soothe the throat pain afterward. Influential researchers had concluded from the fact that many kids' tonsils were constantly swollen that this "focus" of *Streptococcus* infection should simply be junked. Tonsillectomy was good for you and even prevented more serious afflictions. This all turned out to be flat wrong, but not before millions were robbed of a perfectly fine part of their anatomy. And it wasn't even a religious ritual.

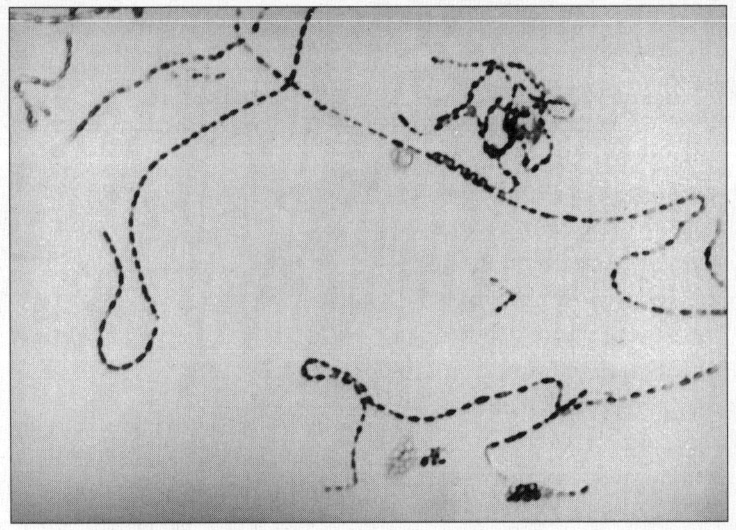

Twisting chains of Streptococcus *bacteria, magnified a thousand times.*

Streptococcus bacteria are admired for growing in beautiful chains; hence the name, first used in 1874, which means a twisted chain of berries. Strep germs are probably most despised today for causing bad sore throats and, occasionally, scarlet fever (which is just sore throat with a red rash and vomiting). They are common throughout the world in school-age children, many of whom are symptomless carriers rather than sick puppies. Incidence of both strep throat and scarlet fever have dropped dramatically since the beginning of this century, perhaps because of decreasing *Streptococcus* virulence and/or increasing human immunity. Of course, antibiotics now keep the bacteria from multiplying in an individual to life-threatening levels.

The first recorded epidemic of scarlet fever, then called *rossalia,* happened in Sicily in 1543. Fluctuating widely in severity, the disease was often confused with measles, as when Samuel Pepys wrote in his diary for November 10, 1664: "My little girle Susan is fallen sick of the meazles, we fear, or, at least, of a scarlett fevour." Scarlet fever often occurred when strep got into the milk supply, perhaps from a dairy farmer with a sore throat, an accident now precluded by pasteurization. (Yet one of the most common microbial causes of harmlessly sour milk is a *Streptococcus.* Another strain gives yogurt a creamy taste.) Before 1830 the disease was apparently mild in America, but deadly epidemics then struck regularly across the country until it was a leading factor of childhood mortality. Fatalities, though not necessarily the number of attacks, began to decline around 1880. By the 1970s, deaths in the United States from once-feared scarlet fever had plummeted from more than six thousand a year in 1910 to a negligible dozen or so. Rheumatic fever, another consequence of strep infection that used to be feared as a predisposer to heart disease later in life, is now rare in the United States, though sporadic outbreaks have been trending upward since 1985. It is still awfully common in poor countries.

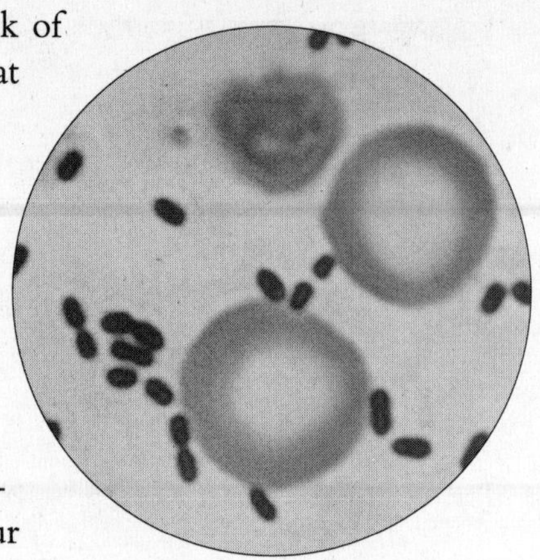

Pneumococcus *cells, aka "Friend of the Aged," magnified a thousand times.*

Crowded living conditions help spread strep through close contact, as classically demonstrated during the American mobilization for World War II, when Navy recruits suffered 59,458 cases of scarlet fever and 43,448 cases of rheumatic fever (other military services had similar outbreaks). During World War I, strep infection was the main complication of influenza leading to fatal pneumonia. Today, *Streptococcus pneumoniae,* also called the pneumococcus, is the most common cause of bacterial pneumonia around the world; it is the germ behind as many as half of all ear infections, the latter being the most frequent reason for taking kids to the doctor in the United States (about 24.5 million miserable trips every year). In the absence of earaches, pediatricians might need to moonlight by driving taxis from time to time, just like history Ph.D.s.

Unlike the ever-mutating *Staphylococcus* bug with which it is often mentioned, strep did not raise serious concern about resistance to antibiotics until the 1970s, when penicillin-resistant strains began to appear. As so-called strep-A bacteria, historically the most common and virulent type in humans, seemed virtually to disappear, strep-B strains rose to the occasion, particularly among infants. Then strep-A reemerged in the late 1980s stronger than ever. A highly resistant *S. pneumoniae* was part of this new microbial offensive, shocking the entertainment world when it suddenly took the life of the beloved "Muppets" creator Jim Henson in May 1990 at the age of fifty. Around the same time, a thirty-year-old Canadian man died six weeks after *S. pneumoniae* infected a splinter cut on his finger. Between the 1940s and 1990s the typical dose of penicillin required to fight strep respiratory infections got bigger by a factor of 2,400, as a result of tougher bacterial resistance. In 1993, pediatricians in Kentucky reported that 28 percent of *S. pneumoniae* cultures from children with acute ear infections were resistant to penicillin.

Strep-A's potential to cause severe skin infections was sensationalized in Great Britain in the spring of 1994 when press reports circulated of a deadly "flesh-eating" bacterium that emits an enzyme able to destroy the fascia, or tissue that binds skin to muscle. The germ's invasive speed is unmatched by any other infectious organism. One Tory MP from Gloucester, where a cluster of cases occurred, called the bug "galloping gangrene that can eat flesh at the rate of an inch an hour." Headlines such as "Public told not to panic as bacterium death toll rises to 12" in the *Times* of London no doubt helped sell the story. Level-headed physicians eventually assured the public that the situation was not unprecedented, and rare in any case. About fifteen hundred to two thousand cases of "necrotizing fasciitis" occur in the United States every year, with fatalities around 30 percent in otherwise healthy adults. In November 1994 the fifty-year-old chancellor of a California community college district died from such an infection five days after complaining of sore throat and leg pains.

Maybe some enterprising editor looked up the term "necrotizing fasciitis" and decided that "flesh-eating" would be more likely to awaken the morning commuters. No doubt the faxes between London and Hollywood were quite active for a while, if not the medical hotlines. Serious streptococcal disease does seem to be on the rise, however, perhaps because invasive strains are circulating against which we have not had time to develop much "herd immunity," as epidemiologists say.

• syphilis •

(Treponema pallidum)

The pox. The serpentine malady. Hot humors from foul women (and vice versa). The French disease. Or, depending on one's national allegiance, the Spanish, Italian, German, Polish, or Portuguese disease. Nobody ever wanted to claim syphilis. Lots of people of every stripe caught it. In 1530, during the era of its first and most virulent attack in Europe, an Italian physician named Girolamo Frascatoro, a classmate of Copernicus's at Padua, published a moralistic poem describing its symptoms and how to treat it. For the sake of metaphor he also told the myth of Syphilus, a shepherd who incurred

Always one of Shorty's girls, never Shorty.

Apollo's wrath and got punished with "foul sores." The tale and the venereal disease resonated deeply with Christian notions of sin, tying poor Syphilus's name forever to the divine retribution that awaits anyone daring to do what comes naturally. If you think Western civilization has advanced past the Middle Ages in this regard, you probably haven't picked up a newspaper or magazine in at least ten years.

Frascatoro was no fool. In hypothesizing that syphilis was caused not by a "mysterious shadow or miasma, not by obstructed humors, but by a kind of seed," he was on the brink of a germ theory of disease.

Treponema pallidum ("pale-turning thread"), the beautiful spiral bacterium that causes syphilis, can be killed in a few minutes by soap and water. It probably cannot penetrate unbroken skin (shaven skin, maybe). It does not like dry places. The disease is thought of as highly contagious, but being exposed to the bug does not always mean infection. Transmission, when it does happen, is usually by hetero- or homosexual intercourse, or heavy kissing, which allows germs to pass from a moist chancre on the already infected partner's genitals or lips to a moist site on the uninfected object of affection. Since the bacteria eventually enter the bloodstream, syphilis can also be passed via transfusions. Penicillin will stop the devil in its tracks, but there is no vaccine: *T. pallidum* is so well adapted to living tissue that it cannot be artificially cultured. Luckily, it has never developed resistance to antibiotics.

Long before these basic facts were collected, syphilis cut a swath through society that was more reprobative than deadly. The disease was first noticed in Europe at the end of the fifteenth century, though it may have existed before then undistinguished from leprosy. In 1494, King Charles VIII of France invaded Italy with Spaniards and other mercenaries. At first there was little fighting and much cavorting with camp followers, resulting in an outbreak of syphilis in Naples that decimated the army (including King Charles) and then spread throughout Europe. Because this initial surge coincided with Columbus's return from the New World, historians have argued ever since about whether *T. pallidum* hitched a ride back inside a few randy sailors. For about fifty years the disease raged with epidemic intensity, affecting as much as 20 percent of urban populations, but then seemed to lose strength and settled into a chronic, degenerative form with fewer fatalities, though still hideous. "A man can no more separate age and covetousness than part young limbs and lechery," Shakespeare wrote in *Henry IV,* "but the gout galls the one, and the pox pinches the other." Its pinched victims included Pope Alexander Borgia, Peter the Great, Benvenuto Cellini ("I believe I caught it from that fine young servant-girl whom I was keeping," he recalled in his autobiography, observing that it covered his body with rosy blisters "the size of six-pence"), Toulouse-Lautrec, Baudelaire, Guy de Maupassant, and Heinrich Heine. Ivan the Terrible's reputation was sealed by the madness of late-stage syphilis. "Everybody has it, more or less," said Flaubert.

By the seventeenth century it was being confused with a lesser demon, gonorrhea. It even gained a reputation for imparting brief periods of feverish creativity, as Thomas Mann wrote in *Doctor Faustus.* For years, mercury was considered the best remedy for syphilis, leading to innumerable cases of poisoning that were as awful as the disease (hence the saying "One night with Venus, a lifetime with Mercury"). For a while, sarsaparilla was thought to be effective, perhaps explaining why otherwise tough cowboys asked for it at the saloon. *T. pallidum* was not identified until 1905, after which various drug therapies radically reduced its incidence in the developed world. The German researcher Paul Ehrlich, searching for an all-purpose bacteria killer, coined the term *Wünder-Bullet* ("magic bullet") to describe his 606th attempt, an arsenic compound known as Salvarsan, which was effective against *T. pallidum* but produced bad side effects. He persevered to create a safer drug, "914" or Neosalvarsan, which helped combat syphilis during World War I. Still, during the 1920s more than nine thousand Americans died of syphilis and sixty thousand babies were born with *T. pallidum* every year.

Like other sexually transmitted diseases, syphilis tends to accompany promiscuity, drug abuse, and poverty. In a ghastly episode that illustrates how groups can become stigmatized by such afflictions, several hundred African-American men suffering from syphilis in Georgia in 1932 were allowed to develop advanced stages of the disease over a period of decades for the sake of supposed research under the auspices of the U.S. Public Health Service. Known as the "Tuskegee experiment," it had little or no scientific value, given that symptoms were already well documented. Penicillin that could have halted the disease was available, of course, but was withheld. Nazi Germany surpassed such medical scandal only in numbers.

By 1970, syphilis was killing only two of every million Americans. But the number of cases had begun to climb by the late 1960s and crack cocaine has fueled an alarming increase since the late 1980s. Condoms are still the second most certain way to avoid syphilis.

One of *T. pallidum*'s close relatives (so close as to be almost identical), *T. pertenue,* causes yaws, a nonvenereal raspberry-like rash encountered mainly in equatorial countries. Unlike with syphilis, there is ancient evidence of yaws in both the Old and New worlds, leading some historians to conclude that yaws was always a village disease while syphilis required the social conditions of urbanization. It was so common among African slaves in America that "yaw houses" were built to isolate them. In the 1950s and '60s, penicillin almost eradicated yaws, but it has come back, especially in West Africa. Millions of people, mostly children, catch it every year.

Another family member, *T. microdentium,* is a harmless inhabitant of our mouths.

• tinea •

Called ring*worm*, tinea is actually a fungus. Such knowledge provides scant comfort.

Tinea capitis, the scalp variety, was epidemic in North America during and after World War II. Grungy movie theater chairs were often responsible for starting an outbreak, when lots of little boys with wiffle haircuts would scrunch down in their seats every Saturday, either picking up or leaving *Microsporum* or *Trichophyton* fungi (long-haired girls were less likely to get infected). In severe, big-city circumstances, doctors prescribed X-ray treatments to kill the fungus. "The rays must make infected hair fall out," intoned the always gung-ho *Time* magazine in December 1943, not seeing the cancer lawsuits on the horizon. Children in Sault Ste. Marie, Ontario, had to wear white cotton bonnets when ringworm took up residence on the heads of 1,363 of the city's 5,712 elementary-school children in 1950. The bonnets were boiled every night by horrified moms. Happily, various ointments and salves have largely chased away the ringworm-of-the-scalp blues since then, though it can still occur where hygiene is poor. In 1992, NCAA wrestlers had to submit to scalp exams before they could compete in championship matches.

Tinea pedis, or athlete's foot, is another well-known *Trichophyton* infection, familiar to millions of television viewers who will never see a real case. Tinea cruris, called "jock itch" by cognoscenti, tends to bother men whose clothes are too tight or who are just plain old fat. Tinea barbae is an unusual infection of beards, often bacterial but sometimes fungal. There are several more kinds of tinea, but isn't this enough for now?

• *Trichinella* •

The intestinal roundworm *Trichinella spiralis* can get into human muscle tissue if we eat uncooked, infected meat, a condition known as trichinosis. Pork is the classic bearer of bad news, though bear and walrus meat are on record as hurting a few adventurous souls. Over a period of four to six weeks after initial exposure, the microscopic creatures go through a life cycle that eventually puts millimeter-long larvae in tongue, eye, pectoral, and other muscles, causing pain, fever, hemorrhaging, and occasionally death. Trichinosis is easily prevented by cooking pork thoroughly. It is uncommon in the United States and almost unheard of in countries such as France, where pigs feed on root vegetables instead of garbage. The biggest contemporary American outbreak, involving ninety cases, took place in 1990 when Southeast Asian immigrants ate raw pork sausage at a wedding reception.

• tuberculosis •

TB is one of the great common denominators of human life—half the world's population is infected—yet is paradoxically one of the most easily combated. In the last two centuries it has killed about a billion people, but never with the obvious force of social upheaval that characterizes other great epidemics like smallpox or cholera. TB now infects eight to ten million people annually around the world and kills two to three million. Three species of mycobacteria can cause it—*Mycobacterium tuberculosis, M. bovis*, and *M. africanum*—and they have probably been at work since the dawn of humankind.

Evidence of "phthisis," as Hippocrates called it, or "consumption" has been found in neolithic (4500 B.C.) skeletons in Germany and has been written about in diverse ancient texts (except those of the Americas), including the Hammurabi Code of circa 2000 B.C. Historians believe the disease evolved in the Middle East about eight thousand years ago and crossed over to humans from cattle. It may have reached the Americas via migration from Asia across the Bering Sea. Recently discovered mummified remains of a woman who died a thousand years ago in Peru show typical lung scars.

For hundreds of years in England, scrofula—now known to be a mycobacterial infection of the glands—was called "the king's evil" and was treated by divine "royal touch" from the monarch himself until the eighteenth century. But tuberculosis (the word "tubercle" means a small protuberance, nodule, or growth) did not become a major problem until the Industrial Revolution created the squalid urban conditions it needs to spread. Already in the late seventeenth century TB was killing one out of every five Londoners, according to the philosopher John Locke. The disease ravaged native Americans with added intensity after exposure to Europeans and the forced change

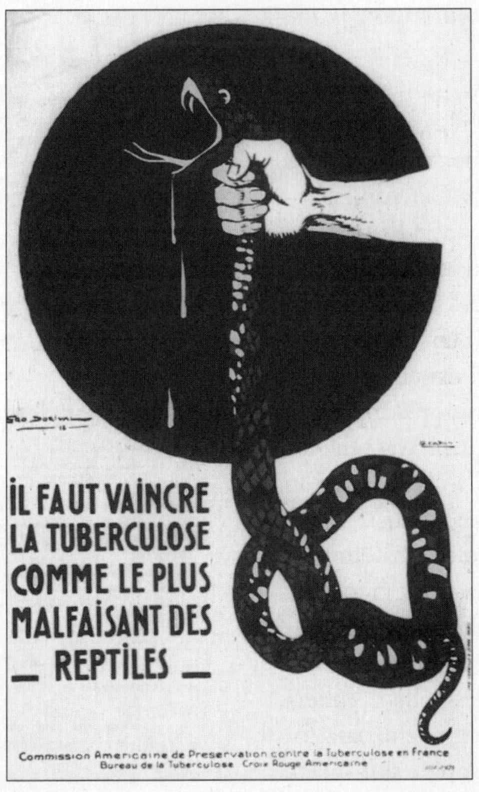

from nomadic to reservation living. René-Théophile-Hyacinthe Laënnec, the inventor of the stethoscope and one of many prodigious tuberculosis researchers in France, died of the disease in 1826. In 1900, TB was killing 200 of every 100,000 (mostly urban) Americans, but as public-health measures improved, the rate fell to 60 per 100,000 by 1940 on the eve of antibiotic therapy.

Coughing, sneezing, talking, singing—anything that makes the bacteria fly out of a carrier's lungs or throat can bring exposure, but infection most often requires months of close contact. Animals also get and pass TB—humans can pick it up from drinking infected cow's milk, though this is rarely a problem where pasteurization is the rule.

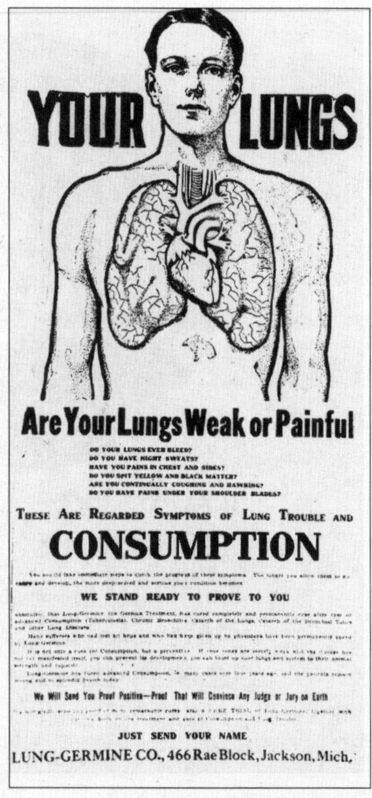

The slow-growing tuberculosis bacillus, *M. tuberculosis*, was identified in 1882 by the great German bacteriologist Robert Koch, who also found the organisms responsible for anthrax and cholera. The object of worshipful veneration in America, Europe, and Japan (where a shrine was built for those who truly believed in his godlike status), Koch took a professional tumble when he later announced a TB vaccine that turned out to be a dangerous dud. Nevertheless, his "tuberculin" did form the basis for today's simple skin-prick

test for tuberculosis—in itself a major development. This test plus chest X rays became the primary tools for tracking the disease.

Until the late 1940s, when the drug streptomycin was found to control it, most sufferers were simply told to rest outside where the air was clean, often at rural hospices designed for that purpose. Galen, the legendary Greek physician of Rome (A.D. 130–200), sent his rich consumptive patients to the beaches of Stabia, across from the Isle of Capri. Thomas Mann wrote about one of the illustrious German sanatoriums in *The Magic Mountain*. TB's victims could constitute a pantheon of the high arts (Keats, Shelley, Chekhov, Chopin, Emily Brontë, Lawrence, Orwell, Stevenson, Whitman), thereby contributing to the romantic notion of creative genius and suffering. "Decay and disease are often beautiful, like the pearly tear of the shellfish and the hectic glow of consumption," Thoreau wrote in 1852. But today more than ever TB is an affliction primarily of the urban poor who do not have access to effective drug regimens. *M. tuberculosis* is a supreme opportunist, waiting for any glitch in the immune system for a chance to strike. Malnutrition is often the door opener, along with chronic infections by other organisms. Since 1985 the number of new cases has risen steadily to about 26,000 a year in the United States, owing to increasing numbers of infected immigrants, homelessness, and the emergence of AIDS. For the period 1986–89, 22 percent of all reported cases occurred in America's foreign-born population (63 percent of these were from Mexico, the Philippines, Vietnam, South Korea, Haiti, or China). For all age groups, tuberculosis is twice as common in blacks as whites, partly because they are less resistant to an initial infection. In 1990, minorities accounted for 70 percent of all TB cases and 86 percent of cases among children.

Compared with some other germs, such as those that cause measles or mumps (where one carrier can infect eighty out of a hundred susceptible people), TB bacteria are rather weak. The disease is thus not highly communicable under decent living conditions—just letting fresh air blow through a room will remove almost two thirds of the infectious material exhaled by a sick person every day. Because the bacteria are sensitive to ultraviolet rays, infection rarely happens outdoors in daylight. Only half of the people who cohabit with someone with infectious TB will likely catch it. But overcrowded American jails are TB hothouses. Prevention is complicated by the fact that at least twelve weeks must pass before a person exposed to the disease will test positive. Moreover, people who test positive may be passive carriers of the disease for years until they develop active TB and thus become infectious to others (some ten million Americans are such carriers). In 1992 an airline flight attendant gave TB to two crew members out of 274 with whom she had worked for varying periods, raising questions about air quality in the confined space of airplane cabins. The two infected persons had spent only about twelve hours with her. In 1994, four travelers on an eight-and-a-half-hour flight from Chicago to Honolulu caught TB by sitting in the same section as an infectious passenger, leading the Centers for Disease Control to advise that people with active TB should not be allowed on commercial flights.

Since treatment is long and expensive, requiring daily drug doses for six to twelve months in conventional cases, shoddy care often results in resistance to the frontline medicines used to fight TB, a serious problem today in the United States. In October 1994 the federal government issued new guidelines for prisons, hospitals, nursing homes, and medical and dental offices to curb the spread of drug-resistant strains. The standards require better ventilation, more efficient masks for workers, and use of ultraviolet light to kill germs. All in all, tuberculosis is an accurate measure of how a society has been doing in the contest between greed and compassion.

Mycobacterium leprae, a cousin of the TB germs, causes leprosy, found most often in Asia and Africa. Since the 1970s, the illness has also been known as Hansen's disease, after Armauer Hansen, who first diagnosed it in detail in 1873. Though the Bible mentions a disfiguring disease assumed to be leprosy, archaeological studies have found no skeletal signs of it before A.D. 500. Leprosy was rampant in Europe from around 1200 until the Black Death era in the fourteenth century, when the sheer loss of human hosts probably stymied *M. leprae* and perhaps opened a niche for *M. tuberculosis*. Some eleven to twelve million cases occur annually around the world today, usually nonfatal but often disabling. Drug-resistant strains began appearing in 1977. As with TB, catching leprosy requires prolonged close contact with an infected person. The germ may reside in armadillos.

• typhoid •

The word "typhoid" means "stuporlike," which describes the mental state of untreated victims. It is caused by a bacterium, *Salmonella typhi*, that comes in more than fifty strains and inhabits only the human digestive system. Though the disease is uncommon now in the United States, its name is firmly entrenched in the lingua franca because of one Mary Malone, a cook who never herself fell ill with the disease but spread it far and wide and became known as Typhoid Mary.

Typhoid is a filth disease usually caught from contaminated food and water. Thus, like typhus, with which it is often confused, typhoid has always tended to appear when hygiene and sanitation break down as a result of war or poverty (though a Stanford University frat house experienced an outbreak in the late sixties when a toilet leaked into the kitchen). During the Boer War, *S. typhi* killed 13,000 British soldiers (compared to 8,000 lost in combat) and sent another 64,000 home early. In the Spanish-American War, 20 percent of U.S. forces got typhoid fever, and there were more than 1,500 deaths. By World War I, the role of infected feces was understood and supplies of fresh water and toilet paper to the troops led to a sharp drop in cases. During this era, typhoid was common in American towns and cities, with especially high death rates in Pittsburgh and Troy, New York.

Because of Typhoid Mary, *S. typhi* is infamous for its ability to reside for long periods of time without producing illness in its host while being able to infect other people and cause typical symptoms. This obviously complicates any effort to control the disease. Mary Malone came to America from Ireland as a teenager and was employed as a cook in well-to-do households. In the winter of 1906–7, an epidemiologist and sanitary engineer named George Soper was hired by the owner of a home in Oyster Bay, New York, to investigate a typhoid outbreak there in a wealthy family that had rented the house during the previous summer. Soper tested the obvious water sources but found no pathogens. He finally interviewed Malone and discovered through her work record that her employment had coincided with seven separate outbreaks of typhoid. When he confronted her with this evidence, she threatened him with a carving fork. She was, after all, in excellent health. Soper then recommended to New York City health department officials that Malone be arrested. This was accomplished after a more violent fork scene with city police, whereupon she was isolated for testing and treatment.

Hearst newspapers soon sensationalized the story of "Typhoid Mary," a sobriquet created by city health officials. After spending two years in a hospital bungalow on North Brother Island, she was released under legal pressure in 1910 on a promise not to resume her trade as a family cook. She promptly adopted aliases and found regular work in hotels and restaurants. Her infective trail ended ironically at the Sloane Hospital for women in 1915, where twenty cases of typhoid were blamed on her kitchen job. She was confined again on North Brother Island until 1932, when a severe stroke sent her to a Bronx hospital, where she died of pneumonia six years later. In all, she was directly linked to at least fifty-three cases of typhoid, three of them fatal.

Today, some four to five hundred cases of typhoid are reported every year in the United States. There are believed to be about two thousand chronic carriers in this country, most of them elderly women, some like Mary Malone with no history of clinical illness. Typhoid used to have a mortality rate around 12 percent before antibiotics, but now deaths occur in fewer than one percent of cases, usually among the very young or old and the malnourished. Clean water, pasteurized milk, and sanitary food handling are the best prevention. The disease is still common in Africa, Southeast Asia, Central and South America, the Middle East, and India—in other words, wherever masses of people live together with little or no sanitation.

• typhus •

"Typhus" refers to two types of fever-rash illness that are rickettsial diseases (*see Rickettsia*). Epidemic typhus, also known as jail fever, is caused by *Rickettsia prowazekii* microbes carried by body lice from person to person. Murine (or endemic) typhus is caused by *Rickettsia typhi* brought to humans from rats by the same flea that can transport bubonic plague (*see Yersinia pestis*). They are both associated with filthy, crowded living conditions.

For centuries, epidemic typhus was a loyal follower of war and famine. The body louse (*Pediculus humanus*, aka "cootie") feeds on infected blood, then jumps to other hosts, where it defecates live *R. prowazekii* organisms that get scratched into the skin. Stopping the spread of disease entails getting rid of the lice. None of this was understood until the first two decades of this century, when the researchers Howard T. Ricketts and Stanislas von Prowazek did pioneering work. Both died of typhus.

The first record of typhus was an outbreak in a Sicilian monastery in 1083. Throughout the Middle Ages, it raged across Europe wherever war and revolution fostered the squalid conditions typhus requires. Explorers and traders took it to the New World, where it helped exterminate native tribes. Columbus caught typhus in 1494. It added to the immeasurable catastrophe of the Thirty Years' War in Germany in the seventeenth century, gutted Napoleon's army and its opponents in Russia (thirty thousand French victims were left to die in Vilna), and arrived in East Coast American cities as "shipboard fever" among immigrants. In 1850, Zachary Taylor is thought to have died of typhus after just sixteen months in the presidency (some historians now believe he might have had Rocky Mountain spotted fever, also caused by a rickettsia germ). Typhus took hundreds of thousands of victims across eastern Europe during World War I. History's worst typhus epidemic occurred in Russia around the time of the Bolshevik revolution—between 1917 and 1921, there may have been 25 million cases and 3 million deaths. The eighteenth-century Russian statesman Potemkin died of typhus.

During and after World War II, typhus returned to strike Italy, Japan, and Korea, but the most wretched scenario unfolded in German concentration camps. When Bergen-Belsen was liberated in the spring of 1945, nearly all of the 55,000 inmates had suffered from typhus over the preceding winter or were currently afflicted, resulting in thousands of deaths (mortality can be as high as 60 percent among older untreated patients). Delousing, which consisted of dusting with DDT, quickly brought the outbreak to an end.

Today, antibiotics easily take care of typhus if used early on. Vaccines are effective but often not available where most needed. Dusting with pesticides to get rid of lice is still practiced in highly infested areas, but being powdered with potent pesticides like DDT, lindane, or malathion is nothing to look forward to except under extreme duress.

The ability of full-blown typhus to make victims disoriented and deranged found its way into the modern Italian language; *tifo* means "typhus," and the verb *tifare* means "to cheer" or "to be a fan." Thus the slogan of the Forza Roma soccer team is *Tifare è un dovere*—literally, to be a fan is a duty.

• varicella-zoster •

Chicken pox, caused by the varicella-zoster virus, is the last holdout of the common childhood diseases. About 90 percent of Americans have suffered through an itchy, grumpy week of it by the time they reach adulthood. For the vast majority it is a largely benign experience, with the greatest social cost calculated from how much work time parents lose while caring for scabby kids at home ($439 million a year). There are some very rare complications, causing as many as 100 deaths among 3.7 million cases annually, but these usually reflect illness in infants or adults.

The varicella-zoster virus, a member of the herpes family, persists for decades in the body after oatmeal baths are just a yucky memory. Contrary to conventional wisdom regarding childhood diseases, the ubiquitous childhood attack of chicken pox does not give permanent immunity. Later bouts are usually just so mild as to be without noticeable symptoms.

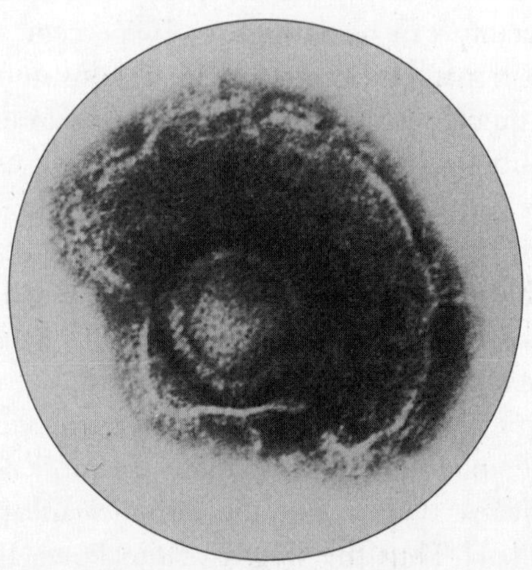

Faithful companion of childhood, the chicken pox virus.

As people pass the age of fifty, however, and their immune systems weaken, the latent virus is more and more likely to trigger a painful skin rash called herpes zoster, or shingles (from the Latin word for belt, *cingulum,* which describes the rash's characteristic band around the ribs). Some 300,000 cases occur yearly in the United States. A study of age incidence has shown that half of people who live to be eighty-five will experience this *recherche du temps perdu,* which is about 25 percent as infectious as chicken pox. Yes, a grandparent with shingles can give chicken pox to previously unexposed kids.

Chicken pox was not recognized as being different from smallpox until the late nineteenth century. In the 1970s, Japanese scientists developed a chicken pox vaccine that has been given to about a million people in Japan and South Korea during the past ten years without incident. A similar shot has been tested on eleven thousand subjects in the United States since 1982 and was submitted for federal approval in 1993; it is already allowed for children and adolescents with leukemia, who are at high risk for serious complications. Critics have worried that routine immunization of children at twelve to eighteen months of age might postpone the disease into adulthood, when it is more hazardous, but experience is quelling this concern. Tests indicate that immunized people are less likely to get shingles. Great caution is taken when introducing such a vaccine because alterations in the universal, intimate relationship between varicella-zoster and humans, the virus's only natural host, could produce new problems.

• worms •

Various nematodes, trematodes, and cestodes—worms to you, bub—cause enormous suffering in poor countries around the world. A combination of climate, unsanitary living conditions, and ignorance puts billions of people at risk for debilitating and potentially lethal disease from microscopic worms.

Guinea worm disease, caused by the nematode worm *Dracunculus medinensis*, comes from contaminated drinking water. Well over a hundred million people are at risk for it in Africa and Asia, but since 1986 the estimated number of cases has fallen from 3.5 million to fewer than 100,000, thanks to a seventy-million-dollar eradication program spearheaded by former President Jimmy Carter. A year after a victim swallows larvae that infect *Cyclops* water fleas, adult worms over one yard long make their way to the skin surface in excruciating, science-fictional horror fashion. For centuries the only treatment has been to wrap the worm's head around a stick and tug it slowly out. This nightmare is now widely avoided by filtering water through special nylon fabric.

Filariasis, caused by parasitic worms in the human lymphatic system and carried in mosquitoes, infects some eighty million people in tropical and subtropical regions of the world. Again, the infection is not fatal, but the presence of adult worms in the lymph system for as long as ten to fifteen years can produce elephantiasis.

Taenia saginata (beef tapeworm) and *Taenia solium* (pork tapeworm) infect humans through undercooked meat. Fifty million people carry both parasites in Latin America, Asia, and Africa, resulting in fifty thousand deaths annually as larvae create cysts in muscle and other tissue.

Onchocerca volvulus causes a disease called river blindness, transmitted by blackflies. Eighteen million people in Africa, Yemen, and Latin America have the worms under their skin, which has led to blindness in some 340,000 cases.

Ascaris lumbricoides, an intestinal worm that causes the disease called roundworm, is staggeringly common in the human race, infecting a billion people in tropical and subtropical countries and killing 20,000 a year through intestinal obstruction. Soil and water contaminated by human feces spread the parasite.

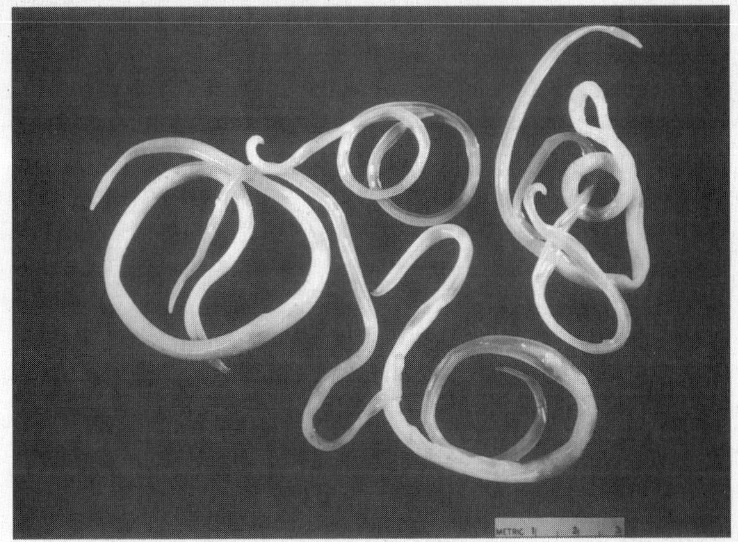

Ascaris lumbricoides, *or roundworm, which can grow to be more than twenty centimeters long, infects a billion people worldwide.*

Ancylostoma duodenale and *Necatur americanus* are two of many parasitic worms that cause hookworm disease in an estimated 900 million people in the tropics, where 60,000 die annually. The worms multiply in the intestines and eventually eat so much blood that the victim gets anemic. Human feces is again the transmitting material.

Parasites in the genus *Schistosoma* infect about 200 million in Africa, Asia, and Latin America through skin contact with water contaminated by human feces or urine. Ever-popular dam and irrigation projects, usually financed by rich countries, have helped spread the problem by widening the habitat of snails that are part of the parasites' life cycle.

Finally, *Enterobius vermicularis*, the pinworm, infects countless humans, especially children, in temperate as well as tropical areas. The parasites live in the intestines for at least three months, then lay eggs near the anus, which prompts itching and scratching. Fingers then go to mouths, where the cycle continues. Pinworms can cause appendicitis and female genital tract problems.

There are more, but these are the big ones. Some might be eradicated with sufficient money and political willpower, others never. They are a constant reminder of worldwide poverty.

• yellow fever •

Yellow fever was the American plague, one of the great tropical fevers that shaped the development of the New World after Columbus. Whether his crew introduced it or not, the yellow fever virus helped exterminate existing civilizations, thus facilitating European conquest, even though whites were as vulnerable as aboriginals. Disastrous urban epidemics in Philadelphia, New Orleans, Charleston, and other blossoming coastal cities tempered nativist optimism about the continent's natural promise. Other diseases may have taken more lives, but none had more cultural or political impact. "Yellow fever will discourage the growth of great cities in our nation," Thomas Jefferson worried in 1800. In fact, it helped drive home the need for basic public services like clean water supplies, sewage systems, and health departments.

The virus is transmitted by certain female *Aedes* mosquitoes, which were not native to the Americas. *A. aegypti*, the classic carrier, likes to lay its eggs in containers of clean water, so it thrives around human settlements, especially in humid climates with an average temperature above 72°F. Epidemics in Barbados in 1647 and in the Yucatan and Cuba in 1648 signaled that both microbe and insect had made the slow voyage from West Africa and established themselves in the Western Hemisphere (the relative immunity of black slaves in the Caribbean suggests yellow fever's origin in Africa). In 1693, Cotton Mather described characteristic symptoms—yellow skin, black vomit—of a disease introduced to

A nativistic nineteenth-century allegorical cartoon showing Columbia rescuing Florida from Yellow Jack, who has apparently sprung from the crate labeled Trade.

Boston by a British vessel from Barbados. Previously unexposed individuals had a fifty-fifty chance of survival. Charleston and Philadelphia were struck for the first time in 1699. New York lost 10 percent of its population in 1702. Transatlantic trade guaranteed that yellow fever would bounce back toward Europe, where the first outbreak occurred in Ca´diz, Spain, in 1730 upon the return of a fleet from Cartegena, Colombia; 2,200 people died.

Philadelphia suffered a paroxysmal outbreak of yellow fever in 1793 that altered the young nation's collective consciousness. Fifteen percent of the city's population succumbed, with more than 4,000 lives lost between August and the arrival of frosty weather in October. People who did not catch the disease chose to flee, virtually emptying the city and prompting New Jersey, Delaware, and Maryland to try to keep out Philadelphians. As during the Black Death of the Middle Ages, the sinews of society were wrenched apart, with family members abandoning each other, officials ordering counterproductive measures, natives blaming foreigners (in this case German and Haitian immigrants), clerics finding reasons not to assist the helpless, and corrupt merchants exploiting everyone. A provisional government was established to maintain some city services in the absence of most office holders. Dr. Benjamin Rush, the country's most eminent physician and a powerful proponent of the "miasma"—that is, environmental—theory of disease, thought for a while that yellow fever stemmed from a load of rotten coffee beans dumped

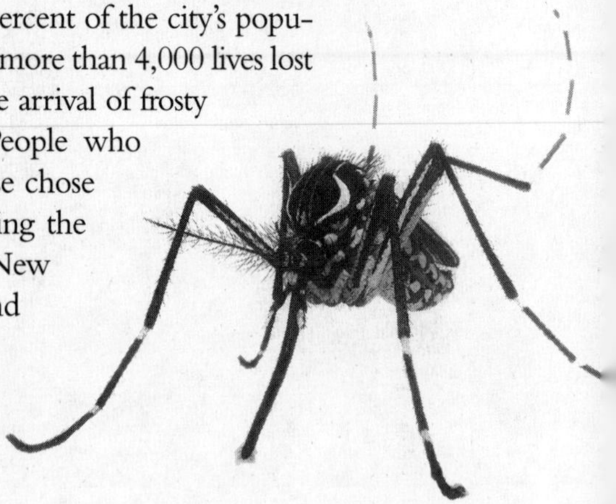

An Aedes aegypti *mosquito (not actual size), first identified as the carrier of yellow fever in 1881.*

on the docks. There was no mass violence, however. Other East Coast cities also experienced serious yellow fever epidemics, and the disease no doubt hastened Philadelphia's slippage behind New York as an important trading port.

Yellow fever had momentous consequences elsewhere in the world, as well. In 1802 Napoleon tried to defeat a slave revolt in Haiti with 25,000 troops, but lost 22,000 of them to yellow fever. The debacle soured Napoleon on the New World, contributing to his sale of the Louisiana Territory in 1803, thus doubling the area of the United States. With no army of occupation, Haiti declared independence in 1804, becoming the first black nation to free itself from European colonial rule.

In Africa, the yellow fever virus was a general in the microbe army that prevented European exploration of the continent until the latter part of the nineteenth century. African slaves were

Devoted scientists examining rats for bubonic plague in New Orleans, 1914.

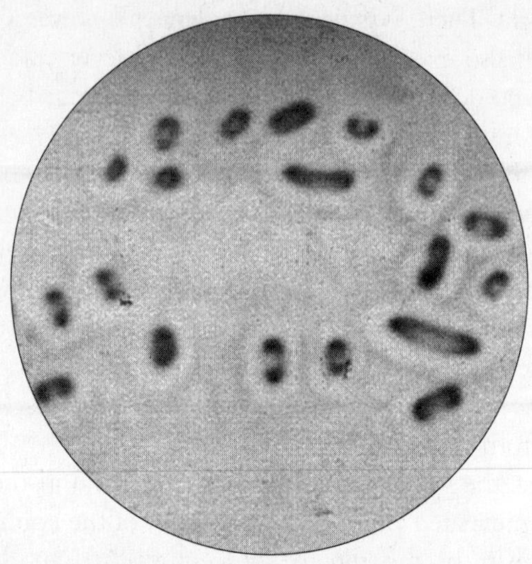

Plague bacteria, once capable of destroying civilization, magnified a thousand times.

known to be relatively immune to yellow fever, malaria, and other diseases that ravaged Europeans and native Americans, though the reasons were of course totally obscure. White surgeons aboard "Guineaman" slave ships had observed these afflictions along the African coast without being able to discern different symptoms. Yellow fever was so common on trade ships between Africa and America that it was dubbed "Yellow Jack" after the flag flown from quarantined vessels. Native Africans incorporated into folk music the role of fever and mosquitoes in protecting them from white colonists.

Until the link between yellow fever and *Aedes aegypti* was proved in 1902, there was no reliable control of the disease. Many observers had drawn the right conclusion empirically; Dr. Carlos Finlay, a Cuban physician, even zeroed in on *Aedes* as early as 1818. The work of Dr. Walter Reed's Yellow Fever Commission, supported by recent proof of mosquito involvement with malaria, finally removed all doubt. Commission members and military volunteers who allowed themselves to be bitten by infected mosquitoes were the heroes of the research. The disease was soon halted in Cuba, Rio de Janeiro, Panama, and Brazil. Ironically, a campaign for worldwide eradication, spearheaded by the Rockefeller Foundation in 1913, met defeat in yellow fever's original breeding ground. In Africa and South America a jungle variety of the disease can pass between monkeys and humans via a number of forest canopy mosquitoes. Breaking this circle of infection was out of the question. A vaccine was developed by 1937 and given on a mass basis beginning in French West Africa in 1939. An epidemic in Ethiopia between 1960 and 1962 infected more than 100,000 people, killing a third of them.

Today, yellow fever occurs mostly in underdeveloped, rural areas where control measures and/or immunization are lax. It still thrives in Africa—an outbreak in Kenya in 1993 underscored its continuing threat—and in South America, where it causes more than ten thousand deaths every year. Mortality can be as high as 10 percent in large outbreaks, though the disease is often mild enough to escape detection.

• *Yersinia pestis* •

The bacterium *Yersinia pestis* was once the planet's most nightmarish life form, making all lions, tigers, and bears seem like kids' stuff. The capital-P Plague, aka Black Death, caused by this bacterium set the physical, social, and metaphorical standards for our perception of epidemic disease. Today it rarely occurs (in the United States, mostly in Arizona, New Mexico, and Colorado), and can be controlled by antibiotics when it does. But in its heyday, *Y. Pestis* more than earned its place as one of the Four Horsemen of the Apocalypse.

Bubonic plague comes from the bites of various types of fleas (principally *Xenopsylla cheopsis,* the Oriental rat flea) that have fed on infected animals. ("Bubonic" refers to the *buboes,* or incredibly painful swollen lymph nodes, that are an early symptom.) The animals include squirrels, marmots, prairie dogs, rabbits, mice, house cats, and, classically, the common domestic rat, *Rattus rattus*. No rats, no fleas, no plague. This relationship, which sounds so straightforward, was not fully explained until 1914. For at least fifteen hundred years, people applied every imaginable combination of folk wisdom, magic, and guesswork to combat the disease. As has often been the case in medicine, the grotesquerie of the "cures" only reflected the desperation of the sick. Drinking witches' brews of snakes and gemstones, inhaling the stench of human feces, building houses to face away from ill winds—when nothing works, anything is worth a try. The nursery rhyme "ring-a-ring o' roses" (or "ring around the rosey") may hark back to the belief that a fragrant bouquet of flowers and herbs would ward off the stinking plague. A few public health innovations, like street cleaning and quarantines (from the Italian: in 1348, Venice ordered all travelers from plague-beset regions to be isolated for forty days, a *quarantena*), were somewhat helpful, but many just made matters worse.

Y. pestis probably moved into its mobile rodent home in primordial Africa or India. *The Iliad* mentions a disease linked with rats during the Trojan War (1190 B.C.) that might have been plague. But the first reliable report was Procopius's description of the so-called plague of Justinian in A.D. 542–43, which struck the Mediterranean from Egypt or Ethiopia. Deaths were said to reach ten thousand a day in Constantinople, with 20 to 25 percent of the region's population wiped out over a five-year period. Some reports claimed that half of the Roman Empire's subjects died, hastening its fall. This was the first of three mighty waves that left clear evidence as *Y. pestis* that moved around the planet with commerce, war, and the bacteria's own wicked pulse.

The second, called the Black Death after the gangrenous extremities of its victims, began in the fourteenth century. It started on the Mongolian steppes and spread westward with commerce across Asia, having already reduced the population of China from 123 million in 1200 to 65 million in 1393. It may ultimately have killed a third of all Europeans (20 million to 30 million people). The infamous story of plague-infested corpses being catapulted over the walls of besieged cities by encircling armies stems from the Tatar attack on Kaffa (now Feodosiya) in the southeast Crimea in 1346. Carried aboard ships by rats and fleas, *Y. pestis* spread around the Mediterranean and throughout Europe during an epoch when medieval political and economic systems were already strained by population growth and other

"The Plague Pit," an early nineteenth-century engraving, depicts victims being dumped from a cart into a mass grave after dark.

social problems, and thus it helped quite literally to clear the way for new ideas. Some of the seminal cities of the budding Renaissance—Florence, Venice, Siena, Orvieto—lost perhaps 50 to 60 percent of their inhabitants. Boccaccio wrote (not as an eyewitness, however) of the destruction of traditional authority at all levels, from family to state. Paris had a death rate of 800 a day. Jews were exterminated as scapegoats to an extent not to be seen again until the Nazi Holocaust. Sixteen thousand were murdered in Strasbourg alone. As late as 1656, Naples lost 300,000 citizens to the disease in five months. In August and September 1665, London lost 7,000 a week to the great plague. "Lord! How sad a sight it is to see the streets empty of people," Samuel Pepys wrote on August 16 of that year. The Great Fire of 1666 helped to end the carnage by destroying rat habitats in thatched roofs. Moscow suffered the final slaughter by *Y. pestis* in Europe in 1771.

The third epidemic started in China during the mid-nineteenth century. Research teams ventured there in the early 1890s, when the French-Swiss bacteriologist Alexandre Yersin discovered the responsible organism in Hong Kong. The germ made its way to the western United States probably around 1900, and the plague ran on through World War II. DDT largely stopped its spread, yet it endures in ground squirrel and other wild rodent burrows. Between 1944 and 1993, in the United States, 362 cases of human plague were reported (10 in 1993). Since the last giant death toll of more than 10 million during the first decade of this century in India, outbreaks have been relatively limited. During the anarchy that followed Indian independence in 1947, 57,000 people died in a single state. Thousands of cases of plague were reported during the war in Vietnam. Today, about 7,000 cases occur worldwide every year.

Although another epidemic of Black Death proportions is extremely unlikely, multiple cases of plague could still appear wherever people live in crowded, unsanitary circumstances near infected rodents. Helter-skelter suburbanization in the southwest has been a prime factor behind recent cases in the United States. An infected rat and two infected fox squirrels (*Sciurus niger*) were found in Dallas in 1993. The mortality rate for untreated bubonic plague is about 60 percent. The single American fatality in 1993 did not receive medical attention until six days after he started feeling ill, at which point he died within six hours of entering the hospital.

Pneumonic plague results when *Y. pestis* enters the lungs via the breath of someone else already infected instead of via a flea bite. Death is virtually certain if treatment is not begun within twenty-four hours, despite the organism's vulnerability to many common antibiotics (except penicillin). In September 1994, more than 400,000 residents of the western Indian industrial city of Surat fled an outbreak of pneumonic plague believed to be the first anywhere in the world for decades. A symbol of squalid commercialization, Surat is known for its pervasive stench arising from refuse, excrement, animal carcasses, and textile-processing chemicals; it produces a thousand tons of garbage every day but disposes of less than half of it. Just one eighth of the city's area has sewage service. During the 1994 outbreak, vans roamed panic-stricken neighborhoods urging the entire population of two million to stay put and take free handouts of the antibiotic tetracycline, which quickly became a black market commodity. Riot police were called upon to enforce an 1897 plague-control law allowing forced hospitalization of victims and control of crowds around hospitals and pharmacies. As in days of old, doctors abandoned their wards to join the refugees. American health officials stepped up surveillance at U.S. airports to monitor any *Y. pestis* carriers from India (none turned up). More than a dozen foreign airlines suspended all flights to the country as cases appeared in other major Indian cities. The official death count

Dream on—an angel tells Charlemagne of a plant that can stop the Plague.

was below 100 out of some 5,000 reported cases, but undoubtedly missed many. By the end of October, a team of American and Russian investigators for the World Health Organization declared India's cities to be plague-free, but recommended that travelers avoid Surat and rural areas near Bombay, where bubonic plague had arisen. All in all, the frightening episode underlined the effect of India's paltry budget for public health, which is proportionately less than that of Bangladesh, Rwanda, Guatemala, or Pakistan.

So don't feed the squirrels.

• Zika fever •

European explorers and colonizers of Africa who managed to survive sundry killer microbes against which they had no natural immunity nonetheless might have felt pale and sweaty, even on good days. Exotic afflictions like this one probably explain why.

An often mild or inapparent disease caused by a *Flavivirus* (an arbovirus genus), Zika fever comes with the scenery in many parts of Africa and Asia. Half of native Africans in some regions have tested positive for antibodies, which means that they have been infected. Yellow fever, dengue, and several varieties of hemorrhagic fever and encephalitis are also brought by flaviviruses, but Zika is evidently not as dangerous as these others to humans. The few diagnosed cases on record have shown the symptoms of fever, chills, headache, nausea, and rash so typical of tropical or subtropical insect-borne illnesses. A closely related virus causes Spondweni fever, named after the South African district. The precise circle of infection for Zika and Spondweni is unknown, but may involve livestock, which have been found to carry antibodies.

The white man perceived Africa as being "dark" for obvious reasons. Part of the great continent's mystery and threat no doubt stemmed from endemic flaviviruses like Zika.

(You'll never see it, but we needed a *Z*.)

• *Notes to Introduction* •

For more detailed bibliographical information on sources cited below, see the Selected Bibliography, p. 179.

1. It is, though. Many of the most common infectious diseases have strong "socioeconomic factors," meaning that poor people catch them more than the better-off. In the United States, this often translates into a racial factor. From cholera in the nineteenth century to AIDS today, the scientific puzzle is thus complicated by social problems.

2. That is, medicine has always been cloistered when there was an economy to support it as such. In the United States before the twentieth century it was a shaky and often shady occupation. See Starr, *The Social Transformation of American Medicine.*

3. Tuchman, *A Distant Mirror,* pp. 106–7.

4. Rosner, *Medicine in the Bible and the Talmud,* pp. 102–7. See also Kraut, *Silent Travelers,* p. 141. The New Testament somewhat reversed this view: "They that be whole need not a physician" (Matthew 9:12).

5. McNeill, *Plagues and Peoples,* pp. 236–37.

6. Brody, *The Healer's Power,* p. 15. "Ante Patientum Latina lingua est,"

advises a medieval medical maxim: "In the presence of the patient, Latin is the language."

7. In her 1978 essay *Illness as Metaphor,* Susan Sontag cited Karl Menninger's observation that "the very word 'cancer' is said to kill some patients" in support of secretiveness on the part of doctors (pp. 6–7). The basic notion is very old. "A physician is only a consoler of the mind," wrote Petronius in his *Satyricon* in about A.D. 50. "Pars sanitatis velle sanari fruit," Seneca wrote around the same time: "It is part of the cure to wish to be cured." A French treatise on surgery from 1316 advises doctors to "keep up the spirits of your patient with the music of the viol and the psaltery, or by forging letters telling of the death of his enemies, or (if he be a cleric) by informing him that he has been made a bishop."

8. It is certainly true that for many centuries charisma and clout were the best thing the profession had to offer. The combination is not without importance today, when more than half of all people seeing a doctor do not have a diagnosable disease and most of those who do either are incurable or will improve by themselves. See Brody, *Healer's Power,* p. 19. The American Medical Association even opposes open access to the National Practitioner Data Bank, which holds information about some 62,000 doctors, dentists, nurses, and other professionals who have been penalized for crimes, mistakes, or incompetence (*The New York Times,* May 29, 1994, sec. 4, p. 4).

9. To be fair, we must acknowledge that medicine is not the only profession to hold such a monopoly and profit from it. Industrial capitalism, revolutions in technology, and urbanization have fueled a broad movement toward specialization and loss of personal autonomy. Still, patients are dependent on physicians in ways that buyers are not normally dependent on sellers. See Starr, *Social Transformation,* pp. 141–42 and 225–26.

10. My use of the term "puritan" does not imply any religious point of view, though Christian fundamentalists, whom some may see as inheritors of the puritan legacy, still seem to have more nerve than others in making their views public.

11. "Do you recall the panic of the institutions of the social body, the doctors and politicians, at the idea of nonlegalized cohabitation or free abortion?" Foucault asked in a 1975 interview. "One needs to

study what kind of body the current society needs." In Foucault, *Power/Knowledge,* pp. 56, 58. See also "Our Bodies Politic," *Washington Post,* August 7, 1994, p. C1.

12. "Any disease that is treated as a mystery and acutely enough feared will be felt to be morally, if not literally, contagious." Sontag, *Illness as Metaphor,* p. 6.

13. Kraut, *Silent Travelers,* pp. 3, 31–49.

14. Brian Pullan, "Plagues and Perceptions of the Poor in Early Modern Italy," in Ranger and Slack, eds., *Epidemics and Ideas,* pp. 101–3.

15. Ranger and Slack, *Epidemics and Ideas,* p. 6.

16. The most sensitive example of this is the social profile of AIDS, which will probably remain largely confined to marginal groups: gay men, intravenous drug users, their sexual partners, and their children. The fact that African-Americans and Hispanics have been struck disproportionately hard further loads the situation.

17. Ranger and Slack, *Epidemics and Ideas,* p. 151. The Islamic belief that plague was a martyrdom contrasts with the Christian insistence that it was a punishment for sins that could be remedied (p. 17). See Sontag, *AIDS and Its Metaphors,* pp. 180–82. "Fear is pain arising from the anticipation of evil," Aristotle, *Rhetoric,* Book II.

18. They are, after all, very small. A baterium may weigh in at 0.00000000001 gram, compared to about 330,000 grams for an average adult human. There is no argument about which creature will have the longest tenure on Earth. There are somewhere between 300,000 and a million species of bacteria and about 5,000 known viruses. Only a small fraction have been scientifically characterized.

19. Between 1928, when Alexander Fleming accidentally discovered that a *Penicillium* mold would kill *Staphylococcus* bacteria, and 1965, when optimism about conquering infectious disease was at its apex, more than 25,000 different antibiotics were developed. About a hundred are in common use.

20. For someone who probably spent most of his time selling buttons and ribbon, Leeuwenhoek (1632–1723) was a remarkable intellec-

tual adventurer. His microscopes, though of course primitive, were hardly crude. The best could magnify 270 times to differentiate objects 0.0014 millimeters apart. Given that he was the first person ever to gaze upon the creepy-crawly world of microorganisms, he was also amazingly brave—he even examined his own diarrheal excrement.

21. Thus, until about a century ago medical practice was based almost entirely on rule of thumb—a dangerous situation, needless to say. The highfalutin manner of the profession's establishment was perhaps akin to Dr. Marvel's bombast in *The Wizard of Oz*. See McGrew, *Encyclopedia of Medical History*, pp. 25–30.

22. Krieg, *Epidemics in the Modern World,* pp. 4, 21–22. Also, McNeill, *Plagues and Peoples*, pp. 9–14. Yellow fever epidemics in colonial Philadelphia were called "Barbados distemper" and, later, "Palatine fever," depending on the origins of recent immigrants. The epidemic of 1793 took on highly specific political overtones when pro-English Federalists held radical Frenchmen from Haiti responsible for the contagion, while liberal Jeffersonian Republicans countered that the fever had domestic origins. Both sides were right, in that the disease required local mosquitoes *and* an initial pool of infected persons. See Kraut, *Silent Travelers,* pp. 25–30. In the case of the precipitous decline of the native population in the Hawaiian Islands after Captain James Cook's arrival in 1778, a disaster of gothic intensity, foreigners' germs were *entirely* to blame. See A. W. Crosby, "Hawaiian Depopulation as a Model for the American Experience," in Ranger and Slack, *Epidemics and Ideas,* Chapter 8.

23. Lederberg et al., *Emerging Infections,* pp. 42–43.

24. Fee and Fox, eds., *aids: The Making of a Chronic Disease,* p. 26.

25. There are lots of them. An average person carries about 100 billion microbes on the outside of his body and 15 trillion inside. A normal bowel movement contains 100 trillion, since we let go of more than we keep.

26. Foucault, *Naissance de la clinique* (*The Birth of the Clinic*), p. 6. More than two millennia earlier, Aeschylus wrote that health and disease are neighbors "with a common wall."

27. It has been estimated that only 5 percent of the world's health research resources are aimed at developing countries, though their people suffer 93 percent of the world population's early mortality. Shulman, *Introduction to Infectious Diseases,* p. 4.

• *Selected Bibliography* •

Berkow, Robert, ed. *The Merck Manual.* 16th ed. Rahway, N.J.: Merck Research Laboratories, 1992.

Bett, Walter R., ed. *The History and Conquest of Common Diseases.* Norman, Okla.: University of Oklahoma Press, 1954.

Brody, Howard. *The Healer's Power.* New Haven, Conn.: Yale University Press, 1992.

Bumgarner, John R. *The Health of the Presidents.* Jefferson, N.C.: McFarland, 1994.

Cartwright, Frederick F. *Disease and History.* New York: Thomas Y. Crowell, 1972.

Centers for Disease Control and Prevention. *Morbidity and Mortality Weekly Report.* Washington, D.C.: U.S. Department of Health and Human Services, issues dated 12/31/93, 1/21/94, 1/28/94, 2/18/94, 3/4/94, 3/11/94, 4/1/94, 4/8/94, 4/15/94, 4/22/94, 5/6/94, 7/1/94, 7/15/94, 8/5/94, 8/12/94, 9/9/94, 9/16/94, 9/30/94, 10/7/94, 10/21/94, 10/28/94.

Crosby, Alfred W. *Ecological Imperialism.* Cambridge: Cambridge University Press, 1986.

Dale, Philip Marshall. *Medical Biographies.* Norman, Okla.: University of Oklahoma Press, 1987.

Delaporte, François. *Disease and Civilization.* Cambridge: MIT Press, 1986.

Dobell, Clifford. *Antony van Leeuwenhoek.* New York: Harcourt Brace, 1932.

Durey, Michael. *The Return of the Plague.* New York: Macmillan/Gill, 1979.

Dutton, Diana B. *Worse Than the Disease.* Cambridge: Cambridge University Press, 1988.

Eberson, Frederick. *Man Against Microbes.* New York: Ronald Press, 1963.

Evans, Alfred S., ed. *Viral Infections of Humans.* 3rd ed. New York: Plenum Medical Book Co., 1989.

Evans, Alfred S., and Harry Feldman, eds. *Bacterial Infections of Humans.*. New York: Plenum Medical Book Co., 1982.

Fee, Elizabeth, and Daniel M. Fox, eds. *AIDS: The Making of a Chronic Disease.* Berkeley: University of California Press, 1992.

Foucault, Michel. *Naissance de la clinique* (*Birth of the Clinic*). Paris: Presses Univesitaires de France, 1963.

———. *Power/Knowledge.* New York: Pantheon Books, 1980.

Garrett, Laurie. *The Coming Plague.* New York: Farrar, Straus and Giroux, 1994.

Gottfried, Robert S. *The Black Death.* New York: Free Press, 1983.

Gregg, Charles T. *Plague!* New York: Charles Scribner's, 1978.

Harden, Victoria A. *Rocky Mountain Spotted Fever.* Baltimore: Johns Hopkins University Press, 1990.

Hill, Justina. *Germs and the Man.* New York: G. P. Putnam's Sons, 1940.

———. *Silent Enemies.* New York: G. P. Putnam's Sons, 1942.

Hoeprich, Paul D. et al., eds. *Infectious Diseases.* 5th ed. Philadelphia: Lippincott, 1994.

Hopkins, Donal R. *Princes and Peasants.* Chicago: University of Chicago Press, 1983.

Joklik, Wolfgang K. et al., eds. *Zinsser Microbiology.* 20th ed. Norwalk, Conn.: Appleton & Lange, 1992.

Kiple, Kenneth F., ed. *The Cambridge World History of Human Disease.* Cambridge: Cambridge University Press, 1993.

Kirp, David L., and Ronald Bayer, eds. *AIDS in the Industrialized Democracies.* New Brunswick, N. J.: Rutgers University Press, 1992.

Kraut, Alan M. *Silent Travelers.* New York: Basic Books, 1994.

Krieg, Joann P. *Epidemics in the Modern World.* New York: Macmillan/Twayne, 1992.

Lederberg, Joshua, Robert Shope, and Stanley C. Oaks, eds. *Emerging Infections.* Washington, D.C.: National Academy Press, 1992.

Levine, Arnold J. *Viruses.* New York: Scientific American Library, 1992.

Longmate, Norman. *King Cholera.* London: Hamish Hamilton, 1966.

Lyons, Albert S., and R. Joseph Petrucelli. *Medicine: An Illustrated History.* New York: Harry N. Abrams, 1987.

McGrew, Roderick E. *Encyclopedia of Medical History.* New York: McGraw-Hill, 1985.

McNeill, William H. *Plagues and Peoples.* Anchor/Doubleday, 1976.

Milne, Lorus, and Margery Milne, eds. *The Audubon Society Field Guide to North American Spiders and Insects.* New York: Knopf, 1986.

Morris, R. J. *Cholera 1832.* New York: Holmes & Meier, 1976.

Morse, Stephen S., ed. *Emerging Viruses.* New York: Oxford University Press, 1993.

Paul, John R. *A History of Poliomyelitis.* New Haven, Conn.: Yale University Press, 1971.

Payer, Lynn. *Disease-Mongers.* New York: John Wiley, 1992.

Postgate, John. *Microbes and Man.* 3rd ed. Cambridge: Cambridge University Press, 1992.

Quétel, Claude. *History of Syphilis.* Baltimore: Johns Hopkins University Press, 1992.

Ranger, Terence, and Paul Slack. *Epidemics and Ideas.* Cambridge: Cambridge University Press, 1992.

Reid, Robert. *Microbes and Men.* New York: Saturday Review Press, 1975.

Rosebury, Theodor. *Life on Man.* New York: Viking Press, 1969.

Rosner, Fred. *Medicine in the Bible and the Talmud.* New York: Yeshiva University Press, 1977.

Shulman, Stanford T. et al., eds. *The Biologic and Clinical Basis of Infectious Diseases.* 4th ed. Philadelphia: W. B. Saunders, 1992.

Siegerist, Henry E. *A History of Medicine.* New York: Oxford University Press, 1951.

Singleton, Paul. *Introduction to Bacteria.* 2nd ed. New York: John Wiley, 1992.

Smith, Jane. *Patenting the Sun.* New York: Anchor/Doubleday, 1991.

Sontag, Susan. *Illness as Metaphor* and *AIDS and Its Metaphors.* New York: Anchor/Doubleday, 1989.

Starr, Paul. *The Social Transformation of American Medicine.* New York: Basic Books, 1982.

Tuchman, Barbara W. *A Distant Mirror.* New York: Ballantine Books, 1979.

Twigg, Graham. *The Black Death.* London: B.T. Batsford, 1984.

Waitzkin, Howard. *The Politics of Medical Encounters.* New Haven, Conn.: Yale University Press, 1991.

Washington Post, *Health*, issues date 12/21/93, 1/4/94, 5/17/94, 6/7/94, 6/14/94, 7/26/94, 9/27/94, 10/4/94, 10/11/94, 10/25/94, 11/1/94, 12/31/94.

Williman, Daniel, ed. *The Black Death.* Binghamton, N.Y.: State University of New York, 1982.

Wilson, Mary E. *A World Guide to Infections.* New York: Oxford University Press, 1991.

Ziegler, Philip. *The Black Death.* Harmondsworth: Pelican Books, 1970.

Zinsser, Hans. *Rats, Lice and History.* New York: Blue Ribbon Books, 1935.

• *Index* •

About the Author

Wayne Biddle is the author of two previous books: *Coming to Terms* (1981), about scientific jargon; and *Barons of the Sky* (1991), a history of the American aerospace weapons business. He is currently at work on a biography of Wernher von Braun. He lives in Washington, D.C., and rural Maryland with his wife and son.